Clinical  gy

T. Wallace MacFarlane
BDS, DDS, FRCPath, FDSRCPS (Glasgow)
Reader in Oral Medicine and Pathology, Glasgow University
Honorary Consultant in Oral Microbiology to Greater Glasgow Health Board

Lakshman P. Samaranayake
BDS, DDS, MIBiol, MRCPath
Senior Lecturer in Oral Medicine and Pathology, Glasgow University
Honorary Consultant in Oral Microbiology to Greater Glasgow Health Board

With a foreword by
Professor J. G. Collee

Wright
London   Boston   Singapore   Sydney   Toronto   Wellington

Wright
is an imprint of Butterworth Scientific

 PART OF REED INTERNATIONAL P.L.C.

All rights reserved. No part of this publication may be reproduced or transmitted in any form or by any means (including photocopying and recording) without the written permission of the copyright holder except in accordance with the provisions of the Copyright Act 1956 (as amended) or under the terms of a licence issued by the Copyright Licensing Agency Ltd, 33–34 Alfred Place, London, England WC1E 7DP. The written permission of the copyright holder must also be obtained before any part of this publication is stored in a retrieval system of any nature. Applications for the copyright holder's written permission to reproduce, transmit or store in a retrieval system any part of this publication should be addressed to the Publishers.

Warning: The doing of an unauthorized act in relation to a copyright work may result in both a civil claim for damages and criminal prosecution.

This book is sold subject to the Standard Conditions of Sale of Net Books and may not be re-sold in the UK below the net price given by the Publishers in their current price list.

First published 1989

© **Butterworth & Co. (Publishers) Ltd, 1989**

British Library Cataloguing-in-Publication Data

MacFarlane, T. Wallace
 Clinical and microbiology.
 1. Man. Oral region. Pathogens: Microorganisms
 I. Title II. Samaranayake, P. Lakshman
 617'.52207

ISBN 0-7236-0934-9

Library of Congress Cataloging-in-Publication Data
MacFarlane, T. Wallace.
 Clinical oral microbiology/T. Wallace MacFarlane,
 Laksham P. Samaranayake.
   p. cm.
 Includes index.
 ISBN 0-7236-0934-9
 1. Mouth–Microbiology. I. Samaranayake, Lakshman P. II. Title.
 [DNLM: 1. Mouth–microbiology. 2. Mouth Diseases–microbiology.
 QW 65 M143c]
 QR47.M323 1989
 617'.522–dc 19
 DNLM/DLC
 for Library of Congress   88-8164
                                                    CIP

Typeset by KEYTEC, Bridport, Dorset
Printed and bound by Hartnolls Ltd, Bodmin, Cornwall

Foreword

by **Professor J. G. Collee** MD, FRCPE, FRCPath,
Professor of Bacteriology, University of Edinburgh

Remarkable advances have been made in clinical microbiology in recent decades – in bacteriology, virology, protozoology and related immunology. Important models of monomicrobial and polymicrobial mechanisms of pathogenicity have been described, so that the multifactorial nature of many infections is appreciated. This has clear relevance to strategies of treatment and to the control and prevention of many infections.

Microbiologists with special interests in many different branches of clinical microbiology and infectious disease owe much to important pioneer work initiated by oral pathologists and dentists. It is a pleasure to make this tribute in a foreword to a new book written by two very productive workers in the field of oral microbiology.

It is increasingly important that dentists understand the pathogenic mechanisms of a widening range of recognized infections if patients are to be treated effectively and if clinical and paraclinical personnel are to be adequately protected. We have been relatively slow to apply some important lessons in clinical practice; further improvement must depend on more effective dissemination of important practical knowledge to our clinicians and practitioners.

Information on infectious diseases of special relevance to dentists, ranging from streptococcal tonsillitis to hepatitis B, is provided in this text. The range is exciting and the amount of detail presented by the authors is impressive. The integrated approach will commend itself to clinical teachers and to students; and practising dentists should welcome the volume as a succinct updated account of new facts and concepts that continue to evolve. The good balance achieved reflects the experience and good sense of the authors who are to be congratulated on making such a significant contribution to our teaching and understanding at a critical stage in the development of the science and practice of clinical oral microbiology.

I wish this book the success that it clearly deserves.

J. G. Collee

Preface

Our main aim in writing this short book was to produce a low cost text which stressed the many links between oral microbiology and clinical dentistry. Previous textbooks have tended to concentrate mainly on the microorganisms involved in oral and dental disease with little attempt to relate these to clinical matters. Not surprisingly, undergraduate students and many dentists fail to see any relevance between oral microbiology and their everyday clinical work. We hope that this book will help to bridge this gap and clarify the important role which oral microbiology plays in dentistry and that it will be read not only by students but also by dental practitioners.

Glasgow, Scotland T.W.MacF.
October 1988 L.P.S.

Acknowledgements

We wish to take this opportunity to thank a number of our colleagues who have assisted us in completing this book, especially Miss Alexandra M. Garven and Mrs Jessie Davis, for typing the manuscript, and the staff of the Dental Illustration Unit, Glasgow Dental Hospital and School. We are indebted to our clinical colleagues, Dr M. A. O. Lewis, Dr P-J. Lamey, Dr J. Rennie, Dr D. R. Sawyer, Dr M. Partridge and Professor D. A. McGowan, who supplied some of the illustrations and clinical photographs, and Dr D. F. Kinane for his helpful advice.

We also wish to thank Glasgow University and the Greater Glasgow Area Health Board for their joint support of the Oral Microbiology Unit over the past 18 years, especially during the periods of financial stringency.

Contents

Foreword v

Preface vii

Acknowledgements ix

1 **Introduction** 1

Part 1 Oral Ecology

2 **Normal oral flora** 7
Study of the oral microbial flora 8

3 **Oral ecosystem and dental plaque** 21
The oral environment 21
Acquisition of the normal oral microflora 26
Dental plaque 27

Part 2 Diseases related to dental plaque

4 **Microbiology of dental caries** 35
Clinical presentation and diagnosis 35
Aetiology 36
Management of dental caries 44
Prevention of dental caries 46

5 **Microbiology of periodontal diseases** 51
Types of periodontal disease 51
Factors involved in periodontal diseases 52
Chronic gingivitis 60
The transitional stage 62
Chronic periodontitis 63

Localized juvenile periodontitis 67
Prepubertal periodontitis 68
Rapidly progressive periodontitis 69
Acute ulcerative gingivitis 70

6 **Dentoalveolar infections** 75
Dentoalveolar abscess 75
Ludwig's angina 82
Periodontal abscess 83
Suppurative osteomyelitis of the jaws 85
Staphylococcal submandibular lymphadenitis in children 88
Cervico facial actinomycosis 89

7 **Microbiology in endodontic therapy** 93
Source and route of infection 93
Pathogenesis of pulp and periapical infections 93
Host defence mechanisms 95
Microbiology 95
Role of microbiology in endodontics 96
Microbiological sampling from root canals 98
Treatment 99

Part 3 Infections involving the oral and perioral tissues

8 **Bacterial infections** 103
Tonsillitis and pharyngitis 103
Gonorrhoea 103
Syphilis 107
Tuberculosis 112
Leprosy 115
Tetanus 119

9 **Fungal infections** 122
Oral candidosis 123
Candidoses confined to the oral and perioral tissues 126
Oral manifestations of mucocutaneous candidoses 134
Oral manifestations of systemic mycoses 137

10 **Viral infections** 140
Interaction of virus and host cells 141
Recovery and immunity to virus infection 142

Herpes virus infections 143
Varicella zoster infection 148
Infectious mononucleosis 151
Cytomegalovirus disease 152
Coxsackie A virus infections 152
Paramyxovirus infections 155
Viruses and cancer 156

11 Salivary gland infections 158
Viral infections of salivary glands 159
Bacterial infections of salivary glands 161
Oral infection consequential to xerostomia 166

12 Infections in medically compromised patients 167
Infective endocarditis 167
Recommendations on antibiotic prophylaxis 174
Prostheses 177
Immunocompromised patients 177
Oral and dental infections in compromised patients 179
Xerostomia and Sjögren's syndrome 182

Part 4 Diagnostic oral microbiology

13 Use of the microbiology laboratory 187
Requesting a microbiology report 187
Specimen collection from oral infections 191
Diagnosis of fungal infections 198
The use of laboratory investigations in the management of antimicrobial therapy 200

14 Antimicrobial chemotherapy 204
Bacteriostatic and bactericidal antimicrobial agents 204
General principles of antimicrobial therapy 205
Antimicrobial prophylaxis 207
Prescribing an antimicrobial agent 208
Pharmacodynamics an antimicrobial agents 209
Failure of antimicrobial therapy 211
Antimicrobials commonly used in dentistry 211

Part 5 Cross-infection in dentistry

15 Viral hepatitis 223
Hepatitis A 225

Hepatitis B   226
Non-A non-B (NANB) hepatitis   236
Delta hepatitis (hepatitis D)   236

16 Human immunodeficiency virus and acquired immune deficiency syndrome   239
Definition of AIDS   239
Epidemiology   240
Human immunodeficiency virus   240
Natural history and clinical features of AIDS   243
Oral manifestations of AIDS   245
Transmission of AIDS   250

17 Cross-infection and sterilization   254
Cross-infection in the dental surgery   254
Prevention of cross-infection   256
Dental treatment of high-risk patients   263

Appendices

1 Code of practice for the use of protective clothing and other procedures associated with the prevention of cross-infection during dental treatment   267
2 Additional procedures available when treating known high-risk infective patients   274

Index   277

Chapter 1
Introduction

Clinical oral microbiology is the study of specimens taken from patients suspected of having infections of the mouth and surrounding tissues. The main purpose of studying such samples is to obtain information which will assist the clinician in arriving at a definitive diagnosis and in giving advice about the management of the patient, especially concerning antimicrobial therapy.

Since many dentists learn about the principles of microbiology during their medical microbiology course, it is important to emphasize that the more common oral and dental infections, e.g. caries and periodontal diseases, are different from almost all infections in other parts of the body. In medical microbiology, infections are often exogenous and caused by a single pathogenic species, whereas endogenous polymicrobial infections are more common in the mouth. As a result, in the investigation of many oral and dental infections, it is necessary to assess both qualitative and quantitative changes in the normal oral microflora, instead of screening for a single non-commensal pathogen, as occurs in the majority of medical investigations.

Microbial infections of the oral and dental tissues are caused by four main groups of microorganisms. Firstly, there are non-specific localized infections caused by bacteria normally found in the mouth, i.e. commensal bacteria, usually in mixed but occasionally in pure culture. It is a characteristic of this group of diseases that they are not usually attributable to any single specific pathogenic agent but are due to the combined activity of a small number of different microbial species, not always the same in every case. Examples of these endogenous infections are dental caries, periodontal diseases, and dentoalveolar infections. The second group of localized infections are caused by specific microorganisms normally present in the oral flora. Examples of these are actinomycosis and candidosis. Thirdly, there are specific systemic infections with associated oral manifestations caused by microorganisms which are not normally present in the oral flora. Examples include bacterial infections, e.g. syphilis, gonorrhoea, tuberculosis, and also a number of viral infections, e.g. herpetic stomatitis, mumps, chickenpox, shingles and herpangina. Fourthly, there are systemic infections without oral symptoms but caused by microorganisms which are members of the oral flora, e.g. infective endocarditis caused by various oral streptococci. Finally, there are a number of systemic infections which have few oral or dental manifestations but, due to the possible risk of cross-infection which may occur during dental treatment, require to be studied by dentists: such infections include hepatitis B and acquired immune deficiency syndrome (AIDS).

A dentist should be able to diagnose, treat or prevent most of the above endogenous infections, as well as some of the exogenous diseases. In addition, he should be competent to assess the possible risks of cross-infection associated with treating a patient infected with a range of specific pathogens. This can be best achieved if the dentist fully understands host–parasite interactions, i.e. the mechanisms whereby microorganisms are transmitted from one patient to another, the way in which they invade the host tissues and cause tissue destruction, and how the host tissues in turn resist the attack of the microorganisms. It is not necessary for the dentist to know in detail the anatomy, physiology or biochemistry of the microorganisms involved in all oral infections, nor to remember precise details of how they are isolated and identified, although he should possess some information of a general nature about these subjects. The dental clinician requires a sound knowledge of the samples required to diagnose oral infections and how to transport these specimens to the laboratory. In addition, he should appreciate the rationale behind the laboratory tests which are used to assess the sensitivity of microorganisms to antibiotics and the clinical significance of these results. Finally, clinicians should be familiar with the concepts of sterility and disinfection and have a good working knowledge of the methods of sterilization of dental instruments and prevention of cross-infection in the dental surgery.

The aim of this book is to give a clear, concise and up-to-date account of the clinical aspects of oral microbiology and it has been assumed that readers will have a reasonable understanding of the principles of microbiology. The book is intended primarily for dental undergraduate students, although it may be useful to dental postgraduate students, as well as general dental practitioners who wish to update or extend their knowledge of oral microbiology. There are a number of areas in the microbiology of the mouth for which complex and often conflicting data are available. When such a problem has been encountered, we have presented a simplified and personal view which inevitably will not be acceptable to all microbiologists.

Oral microbiology has usually been taught as a preclinical or paraclinical offshoot of general or medical microbiology. This approach is not ideal for undergraduate dental students as they have often had difficulty in integrating this knowledge into the various clinical dentistry specialities which are taught in succeeding years. Therefore, we have attempted to follow a more integrated approach to teaching oral microbiology throughout the book and this is particularly reflected in the broad categorization of the text into five major sections, i.e. oral ecology, diseases related to dental plaque, infections of oral and perioral tissues, diagnostic oral microbiology and cross-infection in dentistry. It is hoped that this framework will enable the reader to obtain a clear, overall picture of the subject. While the general contents of the book should help the novice as well as the postgraduate to expand his knowledge of oral microbiology, some sections, such as diagnostic oral microbiology and cross-infection in dentistry, should have particular appeal to the busy general dental practitioner who wishes to use the text as a quick reference manual.

Part 1
Oral ecology

Chapter 2
Normal oral flora

Since the human mouth contains a wide range of sites with different environmental characteristics, it is not surprising that the oral flora consists of a complex mixture of microbial species which include bacteria, mycoplasma, fungi and protozoa. It is not possible in this short book to discuss individual microbial species in detail and, therefore, only the relevant features of the more important members of the oral microflora (selected on the

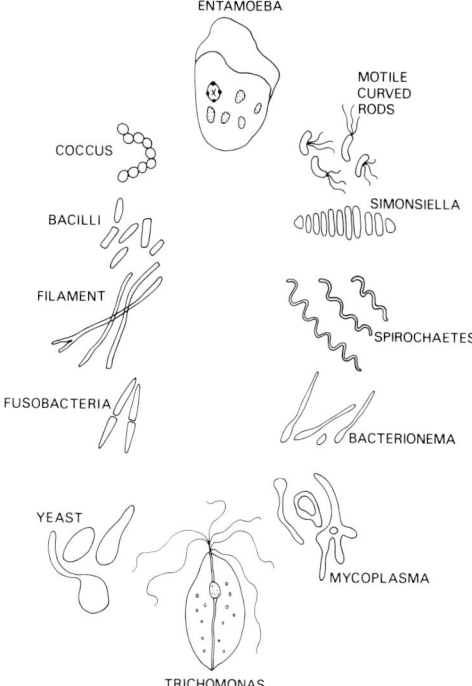

Figure 2.1 The main morphological types of microorganisms present in the human oral cavity

basis of numerical dominance and/or relation to disease) are given. Other microbial characteristics, e.g. factors related to pathogenicity, will be presented in later chapters when specific diseases are discussed. The information is presented in a tabular format (Table 2.1) so that students can refer easily to the appropriate section when they encounter a species with which they are unfamiliar.

The main morphological types of oral microorganisms are shown in Figure 2.1.

Study of the oral microbial flora

The mouth consists of a number of ecological niches which are colonized by a characteristic mixture of microorganisms. The main ecological areas of the mouth are: (1) the mucosa of the lips, cheeks and palate; (2) the tongue; (3) the tooth surfaces (both above and below the gingival margin); (4) saliva; (5) the tonsillar area; and (6) dentures, if present.

Saliva, although widely used in the past as the main sample in studying the oral microbial flora, does not give an accurate qualitative or quantitative assessment of any particular part of the mouth, and is best regarded as representing the overflow of microorganisms from the other oral sites. The problems associated with qualitative and quantitative investigations of the oral microbial flora are discussed below.

Sampling

Saliva
Collect either non-stimulated or wax-stimulated saliva. The results are expressed as the numbers of cultivatable microorganisms/ml saliva.

Mucosal surfaces
Sample standard area with cotton wool swab, polyether foam or by using a gelatin impression technique. The results are expressed as the number of cultivatable microorganisms/mm^2.

Plaque
There are a number of problems, mainly related to the difficulty of access, in sampling plaque. Dental floss, abrasive tape and

Table 2.1 Characteristics of the main microorganisms present in the oral commensal microflora

Genus and main characteristics	Main species	Cultural characteristics	Main intraoral sites/Related infections
Streptococcus Gram-positive cocci in chains, non-motile, usually possess surface fibrils, occasionally capsulate. (Note: The term *Strep. viridans* has been loosely applied as a collective name for α-haemolytic oral streptococci)		Facultative anaerobes, variable haemolysis, but α most common. Selective medium, mitis salivarius agar (MSA)	
	Strep. sanguis	Small rubbery colonies on MSA, firmly attached to agar surface	Mainly dental plaque/Infective endocarditis, dental caries? and aphthous ulceration?
	Strep. oralis	Variable colonial morphology similar to *Strep. sanguis* or small, soft and non-adherent	Tongue, cheek, plaque and saliva/Infective endocarditis
	Strep. mitis	Small non-adherent colonies on MSA	Dental plaque/None
	Strep. mutans	High convex, opaque colonies; extracellular polysaccharide from sucrose. Selective medium MSA + bacitracin; genetically heterogeneous with six species	Tooth surface/Dental caries

Table 2.1 (cont.)

Genus and main characteristics	Main species	Cultural characteristics	Main intraoral sites/ Related infections
	Strep. salivarius	Large mucoid colonies on MSA	Dorsum of tongue and saliva/ None
	Strep. milleri	CO_2 dependent; small non-adherent colonies on MSA. Selective medium contains sulphonamide	Gingival crevice/Dentoalveolar and endodontic infections
Anaerobic streptococcus Small Gram-positive cocci in chains	Peptococcus spp. Peptostreptococcus spp.	Strict anaerobes, slowly growing, usually non-haemolytic	Sub-gingival plaque/ Dentoalveolar infections
Lactobacillus Gram-positive bacilli, catalase negative	L. casei L. fermentum	Microaerophilic; complex nutritional requirements; aciduric; optimal pH 5.5–5.8. Selective medium, Rogosa agar, with low pH due to acetic acid	Dental plaque usually in small numbers/Extension of dental caries especially into dentine
Actinomyces Gram-positive bacilli and filaments. Non-motile	A. israelii A. naeslundii A. viscosus	Microaerophilic strict anaerobe, 'molar' tooth colony on blood agar Facultative anaerobe Facultative anaerobe with requirement for CO_2	Dental plaque and tonsillar crypts/Actinomycosis and dental calculus formation Dental plaque/Root surface caries and calculus formation

Table 2.1 (cont.)

Genus and main characteristics	Main species	Cultural characteristics	Main intraoral sites/ Related infections
Gram-positive pleomorphic bacilli	A. odontolyticus	Facultative anaerobe; reddish-brown centre to colonies	Dental plaque/Extension of caries into dentine
Arachnia Gram-positive pleomorphic bacilli	A. propionica	Strict anaerobe with similar colonial morphology to A. israelii	Dental plaque/Carious dentine, necrotic dental pulp, and chronic periodontitis?
Eubacterium Pleomorphic Gram-positive rods or filaments	E. saburreum E. timidum	Strict anaerobes, characterization ill defined	Dental plaque/Calculus, carious dentine, necrotic pulp and chronic periodontitis?
Bacterionema Gram-positive pleomorphic filaments attached to a rectangular rod shaped body; 'whip handle' appearance	B. matruchotii	Usually facultative anaerobe; some strains strictly anaerobic	Dental plaque/Dental calculus formation
Propionibacteria Gram-positive bacilli	P. acnes	Strict anaerobe, with white colonies surrounded with dark zone on blood agar	Dental plaque/Dentoalveolar infections?
Rothia Gram-positive branching filaments	R. dentocariosa	Usually strict aerobe, with white pigmented colonies	Saliva and dental plaque/None

11

Table 2.1 (cont.)

Genus and main characteristics	Main species	Cultural characteristics	Main intraoral sites/ Related infections
Bifidobacterium Gram-positive bacilli	B. dentium	Strict anaerobe	Dental plaque/None
Micrococcus + Staphylococcus Gram-positive cocci, catalase positive	M. mucilagenosus	Coagulase negative; large colonies adherent to blood agar surface, facultative	Tongue mainly, also gingival crevice/None
	Staph. aureus	Coagulase positive; yellow pigmented colonies, facultative	Saliva of 50% of population in small numbers/Angular cheilitis
Neisseria Gram-negative diplococci	N. lactamicus N. pharyngis	Asaccharolytic and non-polysaccharide producing, facultative	Tongue, saliva, oral mucosa and early plaque/None
Branhamella Gram-negative diplococci	B. catarrhalis	Saccharolytic and polysaccharide producing, facultative	Tongue, saliva and oral mucosa/ None
Veillonella Small Gram-negative cocci	V. alcalescens	Strict anaerobe; selective medium, Rogosa's vancomycin agar	Tongue and saliva and dental plaque/None

Table 2.1 (cont.)

Genus and main characteristics	Main species	Cultural characteristics	Main intraoral sites/ Related infections
Haemophilus Gram-negative cocco-bacilli		All isolates facultative anaerobes; growth enhanced on heated blood agar (chocolate)	Dental plaque, saliva, and oral mucosal surfaces/Occasionally in dentoalveolar infections and acute sialadenitis
	H. parainfluenzae *H. segnis*	V-dependent, non-haemolytic	
	H. aphrophilus	No requirement for V factor	
Actinobacillus Gram-negative cocco-bacilli	*A. actinomycetemcomitans*	No requirement for V factor; 10% CO_2 required for growth. Selective medium contains vancomycin and bacitracin	Occasionally in subgingival plaque/Abscess sometimes with *A. israelii*; juvenile periodontitis
Eikenella Gram-negative cocco-bacilli	*E. corrodens*	X-dependent and microaerophilic; produces corroding colonies on blood agar	Dental plaque/Dentoalveolar abscess; and in some forms of chronic periodontitis?
Capnocytophaga Gram-negative fusiform rods with 'gliding' motility	*C. sputigena* *C. ochracea* *C. gingivalis*	CO_2-dependent (capnophilic), medium-sized colonies with irregular spreading edge	Plaque, mocosal surfaces, saliva/? Destructive periodontal disease

Table 2.1 (cont.)

Genus and main characteristics	Main species	Cultural characteristics	Main intraoral sites/ Related infections
Simonsiella Gram-negative aerobic gliding bacterium with uncommon morphology, i.e. flat, multicellular segmented filaments	*Simonsiella* spp.	Small colonies with narrow band of haemolysis after 24 h at 37 °C on blood agar. For primary isolation use semiselective medium with antibiotics	Dorsum of tongue and hard palate in about 20% of population/None
Fusobacteria Slender, cigar shaped Gram-negative rods with rounded ends	*F. nucleatum*	Strict anaerobe, usually non-haemolytic	Gingival crevice or tonsil/Acute ulcerative gingivitis, dentoalveolar abscess and chronic periodontitis?
Leptotrichia Gram-negative filaments with at least one pointed end	*L. buccalis*	Strict anaerobe with colonies resembling those of fusobacteria	Dental plaque/Not known

Table 2.1 (cont.)

Genus and main characteristics	Main species	Cultural characteristics	Main intraoral sites/ Related infections
Bacteroides Gram-negative pleomorphic rods with rounded ends, non-motile	B. gingivalis B. intermedius	Strict anaerobes, usually require vitamin K and haemin for growth. Black pigmented colonies on blood agar	Gingival crevice and subgingival plaque in small numbers/ Chronic periodontitis and dentoalveolar abscess
	B. melaninogenicus B. endodontalis		Dental plaque/Not known Dental plaque/Root canal and dentoalveolar infections
	B. oralis	Non-pigmented colonies on blood agar	Dental plaque/Dentoalveolar infections
Wolinella Gram-negative curved bacilli motile by polar flagellae	W. recta	Strict anaerobe	Gingival crevice/Destructive periodontal disease?
Treponema Motile Gram-negative helical cells 5–16 μm long with tapering ends into which axial filaments are inserted. Three main sizes: large, medium and small	T. denticola T. macrodentium T. orale T. vincenti	All strict anaerobes, and very difficult to culture. Require enriched media with serum. Characterization poor and confusing	Gingival crevice/Acute ulcerative gingivitis, destructive periodontal disease
Selenomonas Gram-negative curved protozoal cells	S. sputigena	Strict anaerobe	Gingival crevice/Not known

Table 2.1 (cont.)

Genus and main characteristics	Main species	Cultural characteristics	Main intraoral sites/ Related infections
Entamoeba Large motile amoebae about 12 μm in diameter	E. gingivalis	Strict anaerobe; complex medium; cannot be cultured in pure culture	Gingival crevice/Chronic periodontitis?
Trichomonas Flagellate protozoan, about 7.5 μm in diameter	T. tenax	Strict anaerobe; complex medium; difficult to grow in pure culture	Gingival crevice/Not known
Candida Gram-positive budding cells 5 μm in diameter	C. albicans C. glabrata C. tropicalis	Facultative anaerobe; on Sabouraud's medium yields creamy colonies. Only C. albicans germ tube positive	Dorsum of tongue mainly/ Oral candidal infections
Mycoplasma Extreme pleomorphism including cocci, ring, dumb bell and asteroid shapes	M. orale M. salivarium	Minute colonies 0.1–1 mm in diameter, barely visible to the naked eye. Culture on soft nutrient agar with 20% serum, DNA, penicillin and thallous acetate	Dental plaque, gingival crevice/ Not known

specially designed plastic instruments have been used for supragingival plaque, and curettes, paperpoints and washing techniques for subgingival plaque. The results can be reported as the number of cultivatable microorganisms/g wet weight of plaque. However, since the samples are often very small and difficult to weigh accurately, the less satisfactory method of reporting the results for each species isolated as a percentage of the total cultivatable flora has been widely used.

Transportation to the laboratory

To ensure maximal viability of microorganisms, all samples once removed from the mouth should be sent immediately to the laboratory. For the majority of samples from oral infections a transport medium with reducing properties is necessary. There are a wide range of media available; factors which can affect the choice of medium are the sensitivity of the test microorganism(s) to environmental factors, such as pH, temperature and oxygen tension, as well as the time of storage. If there is a considerable time delay between collecting the specimens and culturing them, it may help to store the samples at 4 °C. A typical semi-solid transport medium consists of a balanced salts solution, agar or gelatin, reducing agents, e.g. cysteine hydrochloride, and an indicator of anaerobiosis, e.g. methylene blue.

Dispersion

It is necessary to have a uniform distribution of single microorganisms in the sample to achieve accurate enumeration. All samples from the mouth contain aggregates of bacteria. If too great a force is used to disperse the clumps, microorganisms may be damaged and fail to grow when cultured. If insufficient force is used, a single colony which appears on a culture plate (i.e. a colony-forming unit, CFU) may have arisen from, say, 20 bacteria and not a single organism, as is assumed in colony counting techniques. In both these instances the microbial count would be falsely reduced. The ideal method of dispersion used can only be decided after preliminary experiments. Glass or teflon coated tissue grinders, vortex mixing and low frequency ultrasonic disruption techniques have been commonly used in the dispersion of oral samples.

Culture

The type of culture medium used and the time, temperature and atmospheric composition during incubation depend upon the nature of the microorganisms under study. Variations in these factors can cause profound differences, both in the final bacterial count and the range of organisms isolated. Two main types of culture media have been used.

1. A single highly nutritious medium which supports the growth of most species of bacteria. However, no single culture medium can support the growth of all members of the oral flora but blood agar, supplemented with vitamin K and haemin, comes closest to this ideal.
2. Selective or semi-selective media which permit the growth of specific bacteria in a given sample and inhibit the growth of most other microorganisms (e.g. mitis salivarius agar for streptococci). Careful quality control of selective media is required to avoid the occurrence of false positive or negative results due to errors in the concentration of the inhibitory agents added. In most instances a combination of selective and non-selective media will give the most accurate quantitation of the oral microflora.

Time and temperature of incubation

There is wide variation in the growth rate of microorganisms with the result that different incubation times may be required to produce macroscopic colonies of any particular microorganism. If the incubation time is too short for the organism under study, a false negative result may be recorded and, if too long, the colony may die and prevent biochemical identification being carried out. The incubation temperature for optimal growth of most members of the oral flora is 37 °C.

Atmospheric conditions

The gaseous requirements of oral microorganisms vary considerably. The majority are either strict anaerobes (no growth in the presence of oxygen) or facultative anaerobes (growth occurs both in the presence and absence of oxygen), with strict aerobes (oxygen essential for growth) rarely being found. In addition, a number of species are microaerophilic (requiring small amounts of oxygen), e.g. *Actinomyces* spp., while others are capnophilic having a requirement for 5–10% CO_2, e.g. *Capnocytophaga* spp..

Laboratory technique

The most widely used technique for enumerating the oral flora is to prepare serial dilutions of the original sample (e.g. 10^{-1} to 10^{-8}) in a suitable diluent and then plate out aliquots of the different dilutions on to various solid culture media. The sample is spread evenly over the surface of the agar plate which is then incubated for a predetermined period of time (for plaque samples usually about 5–7 days). Subsequently the plates are examined and the dilution giving about 50 colonies per plate is selected and the number and type of colonies are counted. Ideally each of the colonies should be subcultured to obtain a pure growth and then fully identified, using biochemical and/or serological tests. This is an extremely time-consuming, expensive and difficult task. In addition, the laboratory characterization of a number of oral bacterial species is poorly defined, which partially explains why a percentage of bacteria isolated from many oral samples cannot be identified positively using standard techniques. In recent years the technology of nucleic acid hybridization and monoclonal antibody production has been used to detect and quantify specific bacteria in plaque samples, e.g. deoxyribonucleic acid (DNA) probes for *Actinobacillus actinomycetemcomitans* and *Bacteroides gingivalis*.

Interpretation of results

Due to the many possible sources of inaccuracy which are inherent in the quantitation of the oral microflora, the interpretation of results can be difficult. At best, most studies give an approximation of the bacterial species present in a particular site at a specific time. It is unwise to regard the results as absolute and it is not unusual for different workers to produce conflicting results concerning the composition of the microflora in the same oral site. It is necessary to obtain many samples from one site in the same individual over a period of time before an accurate picture of the flora can be obtained.

Further reading

Grant, D. A., Stern, I. B. and Listgarten, M. A. (1988) *Periodontics*, 6th edn, Mosby, St Louis, pp. 147–197

Hardie, J. (1983) Microbial flora of the oral cavity. In *Oral Microbiology and Infectious Disease* (ed. G. S. Schuster), 2nd student edn, Williams and Wilkins, Baltimore

Maynard, D. and Holt, S. C. (1988) Biology of asaccharolytic black-pigmented *Bacteroides* species. *Microbiological Reviews*, **52**, 134–152

Samaranayake, L. P., MacFarlane, T. W., Lamey, P-J. and Ferguson, M. M. (1986) A comparison of oral rinse and imprint sampling techniques for the detection of yeast, coliform and *Staphylococcus aureus* carriage in the oral cavity. *Journal of Oral Pathology*, **15**, 386–388

Zambon, J. J. (1985) *Actinobacillus actinomycetemcomitans* in human periodontal disease. *Journal of Clinical Periodontology*, **12**, 1–20

Chapter 3
Oral ecosystem and dental plaque

The oral environment

The human mouth is lined by a stratified squamous epithelium which is modified in different areas according to function, e.g. the papillary structure of the tongue mucosa. The oral mucous membrane is interrupted by salivary ducts (both major and minor) and by teeth if present. The gingival tissues are closely bound to the tooth surface by the epithelial attachment, and crevicular fluid passes into the mouth from the gingival crevice. A thin layer of saliva bathes the surface of the mucosa and contained in this layer are epithelial squames, polymorphonuclear leucocytes and members of the oral commensal microflora.

However, within the general environment of the mouth there exists a number of different microenvironments or niches, each of which support their own peculiar microflora (Table 3.1). The microbiological differences between some sites are both qualitative and quantitative, e.g. the flora of the tongue and subgingival plaque, while mainly quantitative differences exist between

Table 3.1 Relative proportions of oral microorganisms in health

Microorganism	Tongue	Saliva	Approximal plaque	Subgingival plaque
Strep. mutans	0	2	2	0
Strep. sanguis	2	2	3	2
Strep. oralis	3	3	3	2
Strep. salivarius	3	3	1	1
Actinomyces spp.	0	2	4	3
Lactobacillus spp.	0	1	1	0
Veillonella spp.	3	3	2	2
Bacteroides spp.	0	1	1	2
Fusobacterium spp.	1	1	1	2
Spirochaetes	0	0	0	1
Candida spp.	3	1	0	0

0, not usually found; 1, occasionally present in small numbers; 2, usually present in small numbers; 3, usually present in moderate numbers; 4, usually present in high numbers.

other surfaces, e.g. tongue and saliva. These variations are due to the complex interactions of a wide range of ecological factors, many of which are poorly understood.

The main factors involved are the following.

Anatomical factors

Under ideal conditions the morphology and spatial arrangement of teeth and their relationship with the gingival tissues tend to prevent food from being forced between them. However, due to untreated malocclusion, irregular teeth and poor restorations, this ideal is rarely attained, with the result that stagnation areas which protect proliferating microorganisms from the host defence mechanisms are usually present. Natural stagnation areas, e.g. occlusal fissures, also occur. Another potential area of weakness in the host defences is the thin, non-keratinized sulcular epithelium which lines the gingival crevice.

Salivary factors

Mixed saliva plays a central role in the microbial ecology of the mouth, receiving secretions from the parotid, submandibular, sublingual and minor glands. The composition of saliva from these glands is different and further variation can be produced by other factors, e.g. increase or decrease in salivary flow rate. The main factors and their effect on the oral microflora are shown in Table 3.2. Although the precise role of these different factors in oral ecology is uncertain, there is good evidence that, overall, saliva is important. Patients with severe xerostomia due to the effect of radiotherapy or degenerative disease of the salivary glands (e.g. Sjögren's syndrome), can develop rapidly progressive dental caries, candidosis and acute sialadenitis (see Chapter 12).

Crevicular fluid

The volume of fluid produced increases with the severity of periodontal disease, being low in health. The composition of crevicular fluid is similar to that of serum and contains a complex mixture of electrolytes (Na^+, K^+, Ca^{2+} and Fl^-) and proteins (including albumin, IgG, IgM, IgA, complement and transferrin) which are part of the host defences. In addition, a variety of enzymes produced both by the host and/or bacteria

Table 3.2 Salivary factors involved in oral ecology

Factor	Mechanism	Effect on oral microflora
Mechanical washing action and swallowing	Powerful mechanism which results in microorganisms either in aggregates or bound to epithelial cells being swallowed and killed in gastric acid	Prevents overgrowth of microflora
Glycoproteins	Important in the formation of acquired pellicle on enamel and in coating epithelial cells	May promote or inhibit microbial adherence to oral and dental surfaces. Nutrient source for microbial growth
Lysozyme, lactoferrin, lactoperoxidase	Inhibit the growth of or kill microorganisms	Most effective against exogenous bacteria but also have some effect on members of the commensal flora
Buffering capacity and pH	Bicarbonate system most important; tends to stabilize pH to about 6.7	Tends to prevent overgrowth of potentially pathogenic miroorganisms which require low or high pH environments for maximal growth
Immunoglobulins	Mainly secretory IgA which coats oral and dental surfaces and interacts with specific antigens if present	May inhibit or in some cases promote microbial adherence to surfaces. Also may alter microbial metabolism

are present, e.g. collagenase, elastase, lysozyme and other proteinases. Polymorphonuclear leucocytes are commonly found in health and, while most are viable and probably have the capacity to carry out phagocytosis within the crevice, they are moribund and non-functional in saliva. In summary, the fluid and its constituents contribute to the defence mechanisms of the mouth, especially in relation to the gingival crevice and the gingival third of the tooth. As a constituent of mixed saliva, the fluid may also contribute to the ecology of more distant oral sites.

Microbial factors

The complex mixture of microorganisms which make up the oral microflora can interact both with each other and with the host factors described earlier. A number of the most important of these interactions are shown in Table 3.3.

Table 3.3 Microbial factors involved in oral ecology

Factor	Effect
Cell wall components, extracellular polymers and surface fibrils	Adherence of microorganisms to oral and dental surfaces. Aggregation of organisms of the same species (homotypic) and of different species (heterotypic). Important mechanisms in plaque formation
Bacteriocins and similar products	Chemicals produced by one microbial species which are antagonistic to cells of the same or related species Also other bacteriocin-like substances which are antagonistic to unrelated species
Metabolic end-products	Antagonistic effects, e.g. organic acids and hydrogen peroxide Nutritional inter-relationships, e.g. extracellular polysaccharides produced by one species utilized by another

Miscellaneous factors

There are a wide range of other factors which, to a greater or lesser extent, can affect the ecology of the mouth. The oxidation–reduction level, or redox potential (E_h), varies in different sites within the oral cavity and is likely to affect the microflora which develops. A low value, e.g. − 100 mV in the depths of dental plaque, is likely to support the growth of strictly anaerobic bacteria, whereas these microorganisms are unlikely to proliferate at a value of + 70 mV which is about the level found in a healthy gingival crevice. Other factors which can disturb the ecological balance are: systemic or topical antimicrobial agents (e.g. antibiotics or antiseptics) which depress the number of some species while allowing others to proliferate; habits such as heavy smoking; or dietary factors, such as the intake of high levels of carbohydrate. In addition, dental treatment, oral hygiene procedures and the onset of oral and dental diseases can result in changes.

Microbial adherence and metabolism

Although the factors involved in the control of the oral microflora have so far been described in specific sections for convenience, it must be stressed that *in vivo* no such divisions exist. For example, the complex mixture of host and microbial factors involved in microbial adherence are shown in Table 3.4. In simple terms adhesion to the tooth surface involves, firstly, a reversible phase due mainly to non-specific, weak electrostatic attractive forces, followed by a more permanent phase involving specific bridging interactions derived both from the host and the commensal flora. Since adherence is a dynamic process, these two phases are continuously occurring. Once the tooth surface has been coated with a layer of microorganisms, bacteria from saliva then adhere to this microbial layer, e.g. *Actinomyces* spp. or *Bacteroides* spp. coaggregating with *Streptococcus sanguis*.

Table 3.4 Factors involved in microbial adherence to oral surfaces

Host factors
Acquired pellicle on enamel (mainly salivary glycoprotein)
Aggregate inducing substances in saliva which can adsorb to tooth surface
Antibodies (IgA) and food lectins which may promote or block adherence
Minerals, e.g. calcium

Microbial factors
Possession of surface fibrils, or fimbriae
Lipoteichoic acid (component of bacterial cell walls)
Extracellular polymers, e.g. polysaccharides which promote adherence to pellicle or to other similar or different microbial species

Given the complexity of the process, there is reasonable correlation between the ability of oral commensal bacteria to

Table 3.5 A comparison of the ability of bacteria to adhere to oral surfaces *in vitro* and their relative proportions *in vivo*

	Ability to adhere in vitro		Relative proportions present in vivo	
	Enamel	Cheek	Enamel	Cheek
Strep. mutans	moderate	low	high–low	low
Strep. sanguis	high	moderate	high	moderate
Strep. salivarius	low	moderate	low	moderate
Actinomyces spp.	low	low	moderate	low
Lactobacillus spp.	low	low	low	low
Bacteroides spp.	low	low	low	low

adhere experimentally to oral surfaces and their actual presence or absence from these same sites using culture techniques (Table 3.5).

Similar complex interactions of host and parasite are also found when the various sources of nutrients for oral commensal growth *in vivo* are considered (Table 3.6). Although there have been many studies of microbial growth *in vitro*, due to experimental and technical difficulties little is known about the growth of oral microorganisms *in vivo*. In addition, the site where a microorganism is growing (e.g. in an occlusal fissure compared to the depths of a periodontal pocket) will have a marked effect on its metabolism.

Table 3.6 Sources of nutrients for microbial growth *in vivo*

Host factors
Dietary factors, e.g. sucrose and starch
Salivary constituents, e.g. glycoproteins, minerals, vitamins and buffering systems
Crevicular exudate, e.g. proteins
Gaseous environment

Microbial factors
Extracellular end-products produced by one microbial species essential for the growth of other species, e.g. isobutyrate by fusobacteria for *T. microdentium*
Intracellular food storage granules

Acquisition of the normal oral microflora

At birth the infant's mouth is usually sterile. However, within a few hours commensal microorganisms from the mother's or nurse's mouth, and to a limited extent some species present in the environment, are able to establish themselves in the infant's oral cavity. The first, or pioneer, species are usually streptococci, especially *Strep. salivarius*. The metabolic activity of the pioneer species alters the environment and provides conditions suitable for the colonization of the mouth by other microbial species. Thus an ecosystem containing a few genera and species becomes more complex step-by-step over a period of time until eventually a stable but highly dynamic system, called a climax community, is reached.

The oral flora of most children by the time of their first birthday contains streptococci, staphylococci, neisseria and veillonella, with the majority also possessing actinomyces, lactobacilli, and fusobacteria. The microorganisms which prefer to colonize the dental hard tissues can usually only be detected in the oral flora when teeth have erupted, e.g. *Strep. sanguis*, *Strep. mutans* and *A. viscosus*. Also related to tooth eruption is an increase in the number of strictly anaerobic species, e.g. anaerobic streptococci and motile Gram-negative rods. On the other hand some anaerobes, e.g. black pigmented bacteroides species and oral spirochaetes, do not appear in significant numbers until adolescence. When the adult dentition is lost the microflora becomes more like that of a child before tooth eruption. However, if dentures are worn then the plaque which accumulates on the acrylic surfaces has similarities to enamel plaque, with resulting alterations in the individual's oral microflora.

Dental plaque

Plaque can be defined as a tenacious microbial deposit which forms on hard surfaces within the mouth and consists of microbial cells and their products, together with host compounds mainly derived from saliva. If plaque develops in the gingival margin area, proteins from the crevicular exudate become incorporated in the plaque. Thus the microorganisms in dental plaque are surrounded by an organic matrix which accounts for about 30% of the total plaque volume and is derived both from the host and from the microbial flora. This matrix acts as a food reserve and as a cement, binding microorganisms both to each other and to various surfaces. It seems unlikely that the matrix acts as a diffusion barrier controlling the entry of substances into plaque and enamel.

In the absence of oral hygiene, plaque can be demonstrated on most dental surfaces but normally it is found in anatomical areas protected from the host defence mechanisms e.g. occlusal fissures, interproximally or around the gingival crevice. Plaque samples are described in relation to their site of origin and can be split into three main groups: supragingival, subgingival and denture-related. Supragingival plaque is commonly subdivided into: fissure (occlusal surfaces), approximal (at contact points of teeth) and smooth surface (e.g. buccal and palatal sites). The microbial composition of these different plaques is shown in Table 3.7. When plaque becomes calcified it is called calculus (see Chapter 5).

Table 3.7 A comparison of the microbial composition of plaques which form in different sites of the human mouth in health

Microorganism	Fissure	Approximal	Subgingival	Denture
Streptococci	4	3	3	4
Actinomyces/lactobacilli	3	4	3	3
Anaerobic Gram-negative bacilli	0	2	3	1
Strep. sanguis	1	3	2	3
Strep. mutans	3	2	0	2
Actinomyces spp.	3	4	3	3
Veillonella spp.	2	2	2	2
Fusobacterium spp.	0	1	2	0
Bacteroides spp.	0	1	2	1
Spirochaetes	0	0	2	0

0, not usually found; 1, occasionally present in small numbers; 2, usually present in small numbers; 3, usually present in moderate numbers; 4, usually present in large numbers.

Formation of dental plaque on a clean surface

Pellicle formation

The surface of a mechanically cleaned tooth becomes coated by a pellicle within a few minutes of exposure to the oral environment. The pellicle consists mainly of glycoproteins derived from saliva. It is important to realize that *in vivo* microorganisms initially attach to the acquired pellicle and not directly to enamel (i.e. hydroxyapatite). It is unfortunate that, in many *in vitro* adhesion studies, saliva has been omitted from the experimental design, which makes extrapolation of the results to the *in vivo* situation difficult.

Deposition of bacteria

The microorganisms which are involved in the development of dental plaque over a 3-week period are shown in Table 3.8. Within 10 hours about 10^4 cells/mm^2 are deposited and these pioneer organisms multiply, producing microcolonies which in time become confluent. During this early period an extracellular matrix develops. This consists of microbial polysaccharides and additional layers of salivary glycoprotein or crevicular fluid components, depending on the site of plaque formation. Accurate figures for the rate of microbial growth *in vivo* are not available. However, it appears that doubling times can vary considerably, both among different bacterial species (being measured in hours rather than minutes), and also between members of the same species, depending on the different intraoral environmental conditions.

Table 3.8 Percentage distribution of microorganisms in smooth-surface dental plaque formed over a 3-week period

Microorganism	6 hours	1 day	2 days	7 days	3 weeks
Gram-positive cocci	79	81	71	60	31
Gram-positive bacilli	8	7	10	20	56
Gram-negative cocci	5	5	11	12	9
Gram-negative bacilli	8	7	8	8	4
Strep. sanguis	9	20	36	35	12
Strep. oralis	62	59	35	21	12
Actinomyces spp.	3	7	7	19	53
Haemophilus spp.	0	3	3	0	0
Veillonella spp.	7	1	5	12	7
Bacteroides spp.	7	4	6	2	2

The factors involved in the development of dental plaque are those described earlier in this chapter (see Tables 3.2 – 3.6). The interaction of these factors leads to the selection of the pioneer microorganisms and their subsequent proliferation. The metabolic products of the pioneer organisms then alter the immediate environment, e.g. create conditions with a low redox potential suitable for anaerobes. Other organisms become incorporated into the plaque with a resultant gradual increase in microbial complexity, biomass and thickness. Eventually the mass of plaque will reach a critical size, when deposition and loss of microorganisms become more or less balanced.

Structure of dental plaque

The structure of 3-week-old dental plaque is shown diagrammatically in Figure 3.1. Plaque consists mainly of cocci, bacilli, filaments (especially in the outer layers) and spiral organisms. Some of the main features of plaque include the following.

1. The bacterial cells near to the enamel surface tend to have a reduced cytoplasm/cell wall ratio, suggesting that they are metabolically inactive.
2. In some areas (especially the outer surface of plaque) cocci attach and grow on the surface of filamentous microorganisms giving a 'corn cob' arrangement.
3. There is a tendency for filamentous bacteria to orientate themselves at right angles to the enamel surface, producing a palisade effect.
4. Within the cytoplasm of some bacteria (mainly cocci) are glycogen-like food storage granules which are available for plaque metabolism when other sources of nutrition are in short supply.

The microbial composition of dental plaque can vary widely within a single individual, for example: (1) at different sites on the same tooth, (2) at the same site on different teeth, and (3)

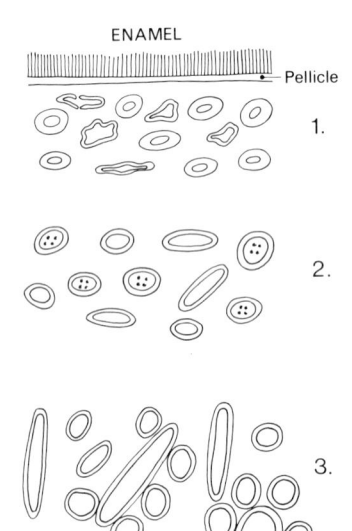

Figure 3.1 The different zones present in mature smooth surface dental plaque. 1, thick-walled inactive bacteria and dead 'ghost cells'; 2, glycogen-like storage granules within bacteria (mainly streptococci); 3, filaments, often at right angles to the enamel surface, with ring of attached cocci ('corn cob' formation)

at different times on the same tooth site. As expected, there are wide differences among individuals and it is important to remember that dental plaque is not a homogeneous material but a complex, dynamic, living biomass. The role of dental plaque in disease is discussed in Chapters 4–7.

Further reading

Carlsson, J. (1986) Metabolic activities of oral bacteria. In *Textbook of Cariology* (eds A. Thylstrup and O. Fejerskov), Munksgaard, Copenhagen, pp. 74–106

Gibbons, R. J. (1984) Microbial ecology–adherent interactions which may affect microbial ecology in the mouth. *Journal of Dental Research*, **63**, 378–385

Liljemark, W. F., Fenner, L. J. and Bloomquist, C. G. (1986) *In vivo* colonization of salivary pellicle by *Haemophilus*, *Actinomyces* and *Streptococcus* species. *Caries Research*, **20**, 481–497

Marsh, P. D. and Martin, M. V. (1984) *Oral Microbiology*, Van Nostrand Reinhold, Wokingham

Nyvad, B. and Fejerskov, O. (1986) Formation, composition and ultrastructure of microbial deposits on the tooth surface. In *Textbook of Cariology* (eds A. Thylstrup and O. Fejerskov), Munksgaard, Copenhagen, pp. 56–73

Socransky S. S. and Manganiello, A. A. (1971) The oral microbiota of man from birth to senility. *Journal of Periodontology*, **42**, 485–494

Part 2
Diseases related to dental plaque

Chapter 4
Microbiology of dental caries

Dental caries can be thought of as a chronic infection in which the microbial agents are members of the normal commensal flora. Lesions result from the demineralization of enamel and later of dentine by acids produced by plaque microorganisms as they metabolize dietary carbohydrates. Once the surface layer of enamel has been lost, the infection invariably progresses via dentine with the pulp becoming firstly inflamed and later necrotic. Dental caries is ubiquitous, and while its prevalence is falling in most developed countries, the opposite is true in the underdeveloped countries where recent changes in diet, especially with respect to carbohydrate content, have occurred.

There is a vast literature about the aetiology, diagnosis, management and prevention of dental caries; in this short chapter only the main factors can be discussed.

Clinical presentation and diagnosis

Dental caries can be classified with respect to the site of the lesion as follows: (1) pit or fissure caries, which occurs on molars, premolars and the lingual surface of maxillary incisors; (2) smooth surface caries, which occurs mainly on approximal tooth surfaces just below the contact point; (3) root surface caries, which occurs on cementum and/or dentine when the root is exposed to the oral environment; and (4) recurrent caries, which is associated with an existing restoration. Depending on the rate of tissue destruction, caries can be described as rampant, slowly progressive or arrested.

The earliest clinical appearance of caries is a well-demarcated chalky-white lesion (Figure 4.1), in which the surface continuity of enamel is still intact. This so called 'white spot' lesion can heal or remineralize with the result that this stage of the disease is reversible. However, as the lesion develops, the surface becomes roughened and cavitation occurs. If the lesion is untreated, microorganisms extend the disease into dentine and often finally destroy the dental pulp.

Diagnosis is usually by a combination of direct observation, probing and radiographs. However, early white spot lesions may easily be missed since they cannot be detected by eye or by radiographs at this stage. It is possible for large carious lesions to develop in pits and fissures with very little clinical evidence of disease. The use of microbiological tests in the assessment of caries will be discussed later in this chapter.

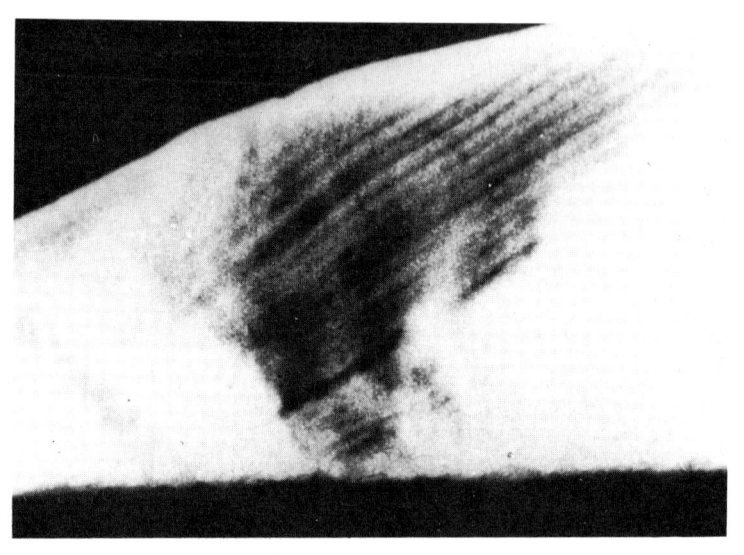

Figure 4.1 A microradiograph of a ground section showing a carious lesion in the enamel. The body of the lesion shows marked radiolucency and the striae of Retzius can be clearly seen. Superficially the surface of the enamel appears well mineralized. (Magnification × 90.) (From Kidd and Joyston-Bechal, 1987, *Essentials of Dental Caries*)

Aetiology

The four main factors involved in dental caries are the host, supragingival plaque, diet and the time necessary for caries development (Figure 4.2). These complex factors can interact in numerous different ways but all are required for the initiation and progression of carious lesions. It is important to realize that the way in which these factors interact is of prime importance in

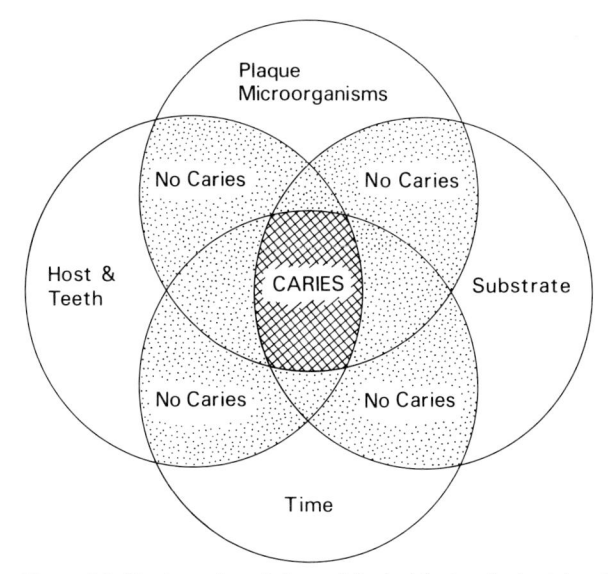

Figure 4.2 The interplay of the aetiological factors in dental caries. All four factors must be acting simultaneously for caries to occur

determining if an early carious lesion will occur and if it will subsequently heal or progress.

Host factors

The two main host factors are the structure of enamel and the composition and flow of saliva. The susceptibility of different areas of enamel on the same tooth to a standard acid attack *in vitro* can vary markedly. This and other information strongly suggest that some areas of the same tooth are much more susceptible to carious attack than others. This fact is not often considered when the role of microorganisms in caries is studied *in vivo* or *in vitro*. Susceptibility to demineralization by acid is related to the mineral content (especially fluoride) and structure of particular areas of enamel.

Saliva performs a number of important roles in maintaining dental health, some which are related to dental caries (see Chapter 3). For example, the mechanical washing action of saliva is a very effective mechanism in removing food debris and unattached oral microorganisms from the mouth. The importance of this factor is highlighted in patients with severe Sjögren's syndrome (a degenerative disease of salivary glands) who have a very low salivary flow rate, retain food debris in their mouth for long periods and suffer from rampant dental caries. Saliva has a high buffering capacity which tends to neutralize acids produced by plaque bacteria on tooth surfaces. It is also supersaturated with calcium and phosphorus which are important in the remineralization of white spot lesions; fluoride is also important in this process.

The roles of the other salivary antimicrobial factors in dental caries, e.g. lysozyme, the lactoperoxidase system and immunoglobulins (see Chapter 3), are not clear.

Diet

A number of epidemiological studies have demonstrated clearly a direct relationship between dental caries and the intake of carbohydrate. The most cariogenic sugar is sucrose and the evidence for its central role in the initiation of dental caries includes the following: (1) increase in the caries prevalence of isolated populations with the introduction of sucrose-rich diets; (2) clinical association studies; (3) short-term experiments in human volunteers using sucrose rinses; and (4) experimental animal studies. In addition, sucrose is highly soluble and diffuses easily into dental plaque, acting as a substrate for the production of extracellular polysaccharide and acids. Although sucrose is not required for the initial attachment of *Strep. mutans* cells to tooth surfaces, it is necessary for subsequent plaque development as an intercellular adhesive. However, the direct relationship between sucrose and dental caries is more complex than can be simply explained by the total amount of sugar consumed. There is good evidence that the frequency of sugar intake, rather than the total sugar consumption, is of decisive importance in caries development. Also important are the stickiness and concentration of sucrose consumed, both factors influencing the period for which sugar is retained in the mouth in close contact with the tooth surface.

Carbohydrates other than sucrose, e.g. glucose and fructose, are also cariogenic but less so than sucrose. Carbohydrates with low cariogenicity also exist, e.g. xylitol (a sugar alcohol).

Microbiology

Dental caries does not occur *in vivo* if microorganisms in the form of dental plaque are absent. The different types of plaque and the factors involved in their development have been described in Chapter 3 and will not be repeated here. Over the years there has been debate about whether one or more specific bacteria are principally involved in the initiation of caries or if the disease is caused by a non-specific mixture of bacteria. At present, a number of different opinions exist, for example: (1) *Strep. mutans* is involved in the initiation of almost all carious lesions in enamel; (2) *Strep. mutans* is important, but not essential, in the aetiology of the disease; and (3) the association of *Strep. mutans* and caries is weak and no greater than for other bacteria in supragingival plaque. The evidence advanced to support or disprove these different opinions is incomplete and it is likely that all three may be correct in specific circumstances. Given the extreme variation found in the microbial composition of supragingival plaque collected from the same site in the same mouth with respect to time, it seems unlikely that the initiation and progression of all carious lesions are associated with identical or even similar plaques. However, there is evidence that overall some bacteria (*Strep. mutans*, *Lactobacillus* spp. and *Actinomyces* spp.) are more important than others.

Streptococcus mutans
A substantial volume of research has been carried out to investigate the role of *Strep. mutans* in dental caries. Surprisingly, few other bacterial species have been investigated in the same depth during the past 10–15 years, and this fact may partly explain the apparently overwhelming evidence of the prime importance of *Strep. mutans* in caries. *Strep. mutans* is a group name; in fact the group consists of six different species (*Strep. mutans*, *Strep. sobrinus*, *Strep. cricetus*, *Strep. ferus*, *Strep. rattus* and *Strep. macacae*) and eight serotypes (a–h). *Strep. mutans* (serotypes c/e/f) and *Strep. sobrinus* (serotypes d/g) are the species most commonly found in humans, with serotype c strains being most frequently isolated, followed by d and e, the others being rarely encountered. The evidence for the aetiological role of *Strep. mutans* in dental caries is shown in Table 4.1.

It must be stressed that not all strains of *Strep. mutans* possess the complete range of properties described in Table 4.1. Therefore, strains of *Strep. mutans* will very likely vary in their potential to produce dental caries. It is possible that certain strains of *Strep. mutans* are more pathogenic than others and that, in a small number of individuals, caries may be an infectious disease, with a highly pathogenic strain being transmitted from one individual to another, e.g. during kissing. However, there is little evidence to support this hypothesis.

Table 4.1 Factors related to the cariogenicity of *Streptococcus mutans*

Significant correlations in humans between *Strep. mutans* counts in saliva and plaque with the prevalence and incidence of caries
Strep. mutans can often be isolated from the tooth surface before the development of caries
Correlation between the progression of carious lesions and *Strep. mutans* counts
Produces extracellular polysaccharide from sucrose which may help to 'cement' the plaque microflora together and to the tooth surface
Most effective streptococcus in experimental caries in animals (rodents and non-human primates)
Ability to initiate and maintain microbial growth and to continue acid production at low pH values
Rapid metabolism of sugars to lactic and other organic acids
Can attain the critical pH for enamel demineralization more rapidly than other common plaque bacteria
Produces intracellular polysaccharide which may act as a food store for use when dietary carbohydrate is low
Immunization of animals with *Strep. mutans* significantly reduces the incidence of caries

Not all evidence supports the apparently strong relationship of *Strep. mutans* and the initiation and progression of caries. Notably, a number of longitudinal studies in children have failed to find any strong correlations. However, there are many problems involved in such studies: for example, the difficulty in diagnosing approximal white spot lesions at an early stage, problems in obtaining plaque samples from the surface of the developing lesion free of surrounding plaque, technical difficulties in the microbiological identification and enumeration of the plaque microflora, and dispute about the best way of statistically analysing the resulting data.

Lactobacilli
Lactobacilli can be divided into two groups: (1) homofermentative species which produce mainly lactic acid (>65%) from glucose fermentation (e.g. *L. acidophilus* and *L. casei*); and (2)

Table 4.2 Factors related to the cariogenicity of *Lactobacillus* species

High numbers in most carious lesions affecting enamel
Numbers in plaque and saliva often positively correlated with caries activity
Some strains produce caries in gnotobiotic rats
Able to initiate and maintain growth at low pH levels
Produce lactic acid in conditions below pH 5.0
Some strains synthesize both extra- and intracellular polysaccharide from sucrose

heterofermentative species which produce lactic acid as well as significant amounts of acetate, ethanol and CO_2 (e.g. *L. fermentum*). The most commonly isolated species from oral samples appear to be *L. casei* and *L. fermentum* although characterization of isolates to species level using existing criteria is not ideal, and not commonly performed.

For many years lactobacilli were believed to be the causative agents of dental caries. Although they possess some properties which would be valuable to a cariogenic organism (Table 4.2), their affinity for the tooth surface and their numbers in dental plaque associated with healthy sites are usually low. In addition, lactobacilli are rarely isolated from plaque prior to the development of caries and are often absent from incipient lesions. On the positive side there is evidence that human lactobacilli strains can cause caries in gnotobiotic rodents. At present, general opinion supports the concept that lactobacilli are not usually involved in the initiation of dental caries but more in the progression of the lesion deep into enamel. Lactobacilli are the main pioneer organisms in the advancing front of the carious process into dentine.

There is evidence from *in vivo* human studies that lactobacilli are closely related to the development of caries in some sites. Generally, good correlations have been described between salivary lactobacillus counts and caries prevalence, although correlations with dietary carbohydrate consumption are less certain. However, few in-depth clinical or laboratory investigations have been performed in recent years and until more information is available the role of lactobacilli in the initiation of dental caries in at least some cases cannot be totally discounted.

Actinomyces
Actinomyces viscosus has been associated with the development of root surface caries. The lesions are different from enamel caries in that the calcified tissues are softened without obvious cavitation. They tend to form on the buccal and lingual surfaces close to the gingival margin and slowly progress laterally around the neck of the tooth. Few detailed *in vivo* microbiological studies of root surface caries have been reported, the majority being cross-sectional rather than the more appropriate longitudinal type. The evidence for the involvement of *A. viscosus* in root surface caries is based on (1) association studies *in vivo*, (2) experimental work with pure cultures *in vitro*, and (3) experimental work in gnotobiotic rodents. While there is little

doubt that *Actinomyces* spp. (especially *A. viscosus*) predominate in the majority of plaque samples taken from root surface lesions, some studies have reported that *Strep. mutans* and *Lactobacillus* spp. were also found in many samples. Furthermore, the sites from which *Strep. mutans* and lactobacilli were isolated appeared to have a higher risk of developing root surface caries than other sites. Factors such as the role of diet, and salivary flow rate and constituents, have received little or no detailed study in this form of dental decay. Therefore, due to the incomplete and often superficial nature of the data available and lack of suitable control samples, it is difficult to be certain if *A. viscosus* is specifically involved in root surface caries or not. This matter will only be resolved when further detailed experimental work is carried out.

Veillonella

There is some evidence which suggests that *Veillonella* spp., which are present in significant numbers in most supragingival plaque samples, may have a protective effect on dental caries. *Veillonella* require lactate for growth but are unable to metabolize normal dietary carbohydrates. Thus, they use lactate produced by other microorganisms and convert it into a range of weaker and probably less cariogenic organic acids, e.g. propionic acid. This protective effect has been demonstrated *in vitro* and in animal experiments but has not been described in humans. However, the mechanism supports the concept of metabolic inter-relationships between dissimilar plaque microorganisms although their importance in modulating the host–parasite balance in enamel demineralization remains uncertain.

Non-specific plaque hypothesis

Although strong evidence exists which suggests that *Strep. mutans* and *Lactobacillus* spp. are commonly associated with the initiation and development of caries, other plaque bacteria also possess some of the biochemical characteristics thought to be important in pathogencity (Table 4.3). Therefore it seems likely that combinations of bacteria other than *Strep. mutans* and *Lactobacillus* spp. may be able to initiate carious lesions.

Plaque metabolism

Saliva is the main source of nutrition for oral microorganisms. The carbohydrate content of saliva is normally low but levels

Table 4.3 Factors which support the non-specific plaque hypothesis of dental caries (data for *Strep. mutans* and lactobacilli excluded)

Streptococci and *Actinomyces* are acidogenic and commonly present in plaque
The total acidogenic count of plaque microorganisms is higher in caries-active than in caries-inactive patients
Streptococci and *Actinomyces* can produce extracellular[a] and intracellular[b] polysaccharides from sucrose
Streptococci and *Actinomyces* are capable of producing caries in gnotobiotic rats[c]

[a] *Strep. sanguis, Strep. mutans, A. viscosus* and *A. naeslundii*
[b] *Strep. sanguis, Strep. mitis* and *A. viscosus*
[c] *Strep. sanguis, Strep. salivarius, Strep. mitior, A. viscosus* and *A. naeslundii*.

can increase 1000-fold following a meal. In order to avoid possible toxic effects and to gain maximum benefit from these high levels of carbohydrate, oral bacteria have developed a number of regulatory mechanisms which act at three main levels: (1) transport of sugar into the organisms; (2) the glycolytic pathway; and (3) the conversion of pyruvate into metabolic end-products. The bacterial metabolism of carbohydrate is important in the aetiology of caries since the end-products are responsible for enamel demineralization.

In most bacteria, glucose is degraded by glycolytic enzymes via the Embden–Myerhof pathway, with the production of two pyruvate molecules from each molecule of glucose. The pyruvate can be degraded further in a number of ways: for example, under low sugar conditions pyruvate is converted into ethanol, acetate and formate, while in sugar excess many bacteria (streptococci and lactobacilli) prevent the accumulation of toxic intermediate metabolites in the cell by increasing their glycolytic rate with more rapid drainage of products from the cell. Pyruvate is converted into lactate molecules. Overall the interaction of plaque bacteria and carbohydrate results in a rapid fall in plaque pH, followed by a slow return to the original value in about one hour, thereby producing a 'Stephan curve'. A range of acids are produced following sugar fermentation, all of which may contribute to the pH responses. In a low sugar environment acetic acid predominates and lactate is low, but when plaque is exposed to carbohydrate a significant increase in lactate occurs, often accompanied by a fall in acetate. In addition, small amounts of butyric, formic, propionic and succinic acid may be present. Since lactate is both the strongest and predominant acid produced during plaque metabolism of carbohydrate, it seems likely that it is the most important in caries development. However, this is not certain and there is some evidence that at low pH levels acetate can preferentially attack enamel crystals.

As a result of the problems associated with (1) the diagnosis of early carious lesions, (2) accurate sampling of demineralized sites, and (3) laboratory difficulties in the isolation, identification and enumeration of plaque microflora and subsequent statistical analysis of the data, together with other associated factors such as enamel susceptibility, diet and salivary characteristics, the precise relationship of bacteria to the initiation of caries remains uncertain.

Management of dental caries

In the past, the general approach in the treatment of dental caries was to remove diseased tissue and replace it with an inert restoration. This form of management made no attempt to cure the disease and the patient often returned some months later requiring further fillings due to new or recurrent caries.

The modern philosophy in dental caries management highlights the importance of accurate diagnosis, minimal cavity preparation techniques, and active prevention. The end result of such measures should be that, with the passage of time, less rather than more restorative work will be required by an individual patient.

Evaluation of the patient

In patients with a low incidence of caries, a case history and clinical and radiographical examination are probably sufficient for treatment planning. However, for patients with rampant or recurrent caries, or where expensive crown and bridge work is planned, additional investigations are necessary, for example (1) assessment of dietary habits, (2) determination of salivary flow rate and buffering capacity, and (3) microbiological tests. It is outside the scope of this book to discuss in detail how these different factors involved in assessing caries risk are performed and subsequently correlated for use in treatment planning. However, the microbiological tests are described briefly below.

Microbiological tests in caries assessment

Saliva samples are used to enumerate the numbers of *Strep.*

mutans and *Lactobacillus* spp. in the mouths of patients. Briefly, a paraffin wax-stimulated sample of mixed saliva is collected and sent to the laboratory, where it is vortex mixed, diluted, and cultured on selective media for *Strep. mutans* (mitis salivarius bacitracin agar) and *Lactobacillus* spp. (Rogosa SL agar). The number of typical colonies at a suitable dilution is recorded and the count/ml saliva is calculated using simple arithmetic. The salivary counts/ml which are accepted as high and low values are: high, > 1 million *Strep. mutans* > 100 000 *Lactobacillus* spp.; low, < 100,000 *Strep. mutans* < 1000 *Lactobacillus* spp.

Generally, salivary counts of these bacterial species correlate well with plaque counts in the same patient. If tests require to be carried out in very young children who have difficulty in producing a stimulated salivary sample, or in patients with severe xerostomia, alternative methods of sample collection are necessary, e.g. using a wooden spatula or a bacteriological wire loop.

A commercially available dip-slide kit (Figure 4.3) is available for lactobacillus counts; it correlates well with laboratory plate counts and can be performed in dental practice without the need for special facilities. A similar kit for estimating salivary counts of *Strep. mutans* has recently become available but the interpretation of results can be difficult.

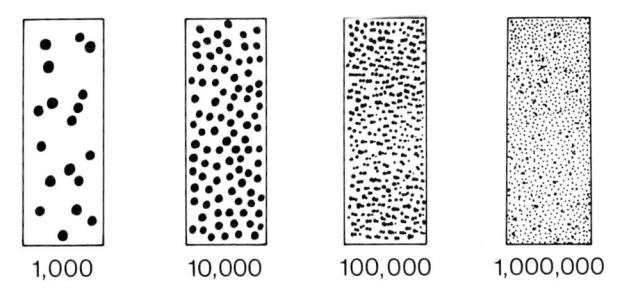

Figure 4.3 The different colony counts which may be obtained on a dip slide for estimating salivary lactobacillus counts

While there are good correlations between caries prevalence and increment and high counts of lactobacilli and *Strep. mutans* when large groups are studied, these tests are accurate in only about 45% of cases when the results for individuals are studied. The presence of high salivary levels of *Strep. mutans* or lactobacilli does not necessarily mean that the patient has an increased risk of developing dental caries because other factors, such as diet, buffering capacity, fluoride content of enamel, and degree of oral hygiene, may combine to produce a protective effect, thus tipping the host–parasite balance from disease towards health. It is important to realize that the microbiological tests used in caries assessment are different from conventional tests used in medical microbiology where the presence of a pathogen indicates a positive diagnosis, e.g. in tuberculosis or gonorrhoea. The main use of microbiology in caries assessment tests is to highlight patients who have abnormally high oral counts of potential pathogens, so that this data can be taken into account when integrating all the factors which may contribute to the carious process in an individual patient. Furthermore, the tests can subsequently be used to monitor the efficacy of preventative techniques, e.g. dietary and oral hygiene advice and the use of antimicrobial agents such as chlorhexidine.

Prevention of dental caries

The four most common approaches used in caries prevention are:

1. To cease between-meal sucrose consumption of carbohydrate or to substitute non-cariogenic sweetners, e.g. sorbitol, xylitol or lycasin.
2. To make the tooth structure less soluble to acid attack by using fluorides.
3. To reduce, if not eliminate, cariogenic microorganisms so that even in the presence of sucrose, acid production will be minimal, e.g. the use of oral hygiene aids, antimicrobial agents and possibly immunization.
4. To use sealants to protect areas of the tooth (e.g. pits and fissures) which are susceptible to carious attack but which cannot easily be kept plaque free using normal oral hygiene procedures.

The rationale for the use of the different procedures is given below.

Sugar substitutes

The use of artificial sweeteners is based on the premise that they cannot be absorbed and metabolized by plaque bacteria to produce acid. Two types of sugar substitutes are available: those which have a calorific value (nutritive sweeteners), e.g. the sugar alcohols, sorbitol and xylitol, and lycasin prepared from corn starch syrup; and non-nutritive sweeteners, e.g. saccharin and aspartame. The majority of oral bacteria cannot metabolize the nutritive sweetners (the possible exception being *Strep. mutans* and sorbitol), and their use appears to induce a degree of remineralization of enamel.

Fluoride and tooth solubility

Fluoride can be administered systemically and become incorporated into enamel during its formation. The best method is via the water supply (1 p.p.m.) but, if this is not possible, tablets or fluoridated milk may be used. Topical application of fluoride can also be used, e.g. as a gel which is applied to teeth in specially constructed trays, or in fluoridated toothpaste. Fluoride ions replace hydroxyl groups in hydroxyapatite and form fluoroapatite which is less soluble in acid. Fluoride also tends to promote remineralization of early enamel and dentine carious lesions.

Control of plaque microflora

A number of different methods can be used to reduce significantly the numbers of supragingival plaque bacteria.

Mechanical cleansing techniques
Conventional toothbrushing with a fluoridated toothpaste on its own may not be very successful in reducing the incidence of caries, partly because it depends entirely on the motivation and skill of the patient. Other aids for plaque removal, e.g. interdental brushes and wood sticks and dental floss, may achieve some reduction in interdental caries, but there is no good evidence for this. Certainly it is unlikely that mechanical cleansing methods will reduce or prevent caries in fissures or pits.

Antimicrobial agents
A wide range of antimicrobial agents have been tried in plaque control but chlorhexidine as a 0.2% mouthwash is by far the most effective. Chlorhexidine acts against many Gram-positive and Gram-negative oral bacteria by damaging their cell membranes. When used as a mouthwash, the antiseptic becomes bound to oral surfaces (especially teeth) and is then slowly released into saliva. This property is unique to chlorhexidine

and to a great extent explains why it is still the most effective oral antiseptic available. The precise mechanisms by which chlorhexidine reduces plaque accumulation, and perhaps also subsequent carious lesion production, are uncertain. However, inhibitory effects on the cell membranes and metabolism of both Gram-positive and Gram-negative oral bacteria have been described. It is also possible that the presence of chlorhexidine on enamel surfaces and in saliva interferes with the adherence of plaque-forming bacteria, thus reducing the rate of plaque accumulation. It is interesting that *Strep. mutans* tends to be more sensitive to chlorhexidine than other streptococci commonly involved in plaque development, e.g. *Strep. sanguis*. This fact is convenient in that the antiseptic not only reduces the numbers and metabolic end-products of the major cariogenic organism but also tends to favour the growth of streptococci with lower cariogenicity. However, due to the dual problems of tooth staining and unpleasant taste, chlorhexidine in normally used only for short-term therapy. In a small number of individuals parotid gland swelling may occur with chlorhexidine rinses. The antiseptic can also be applied to the teeth in gel form contained in special trays.

While there is evidence that systemic antibiotics, e.g. penicillin, tetracycline and erythromycin, and topical agents, e.g. kanamycin, can reduce plaque formation during the period when the drugs are being administered, it is extremely doubtful if there are any situations where their use for caries control is indicated. It has been suggested that antimicrobial agents which are not required for control of serious infections could be used in the dental management of the mentally or physically handicapped. However, the rationale for this is unclear when other local preventive techniques can be used.

In vitro experiments have shown evidence that fluoride may have an inhibitory effect on plaque microorganisms by a number of different mechanisms, for example (1) reduced glycolysis, (2) inactivation of key metabolic enzymes, (3) interference with bacterial membrane permeability, and (4) inhibition of the synethesis of intracellular polysaccharide material. However, the influence of fluoride on the plaque microflora *in vivo* has rarely been studied and it is not clear if all or any of the above mechanisms are active in human supragingival plaque.

Immunization procedures
It is well established that immunization using either cell wall associated antigens or glucosyltransferases from *Strep. mutans*

are effective in reducing experimental dental caries in rats and monkeys. Glucosyltransferases are a mixture of microbial extracellular enzymes which are involved in polymer synthesis from sucrose and thus in plaque accumulation. Vaccines using these enzymes appear to be most effective in reducing smooth surface caries (especially in rodents). A number of cell wall associated vaccines have been tested and all have produced good protection against caries in monkeys. To date, vaccination has not been tested on humans, mainly because of fears related to possible side-effects which would be unacceptable since caries is not directly life threatening. The antibodies which develop after immunization with most *Strep. mutans* antigens tend to cross-react with heart tissue. While the significance of this cross-reaction is unknown, the possibility that heart damage could result must be seriously considered.

Another problem in instituting a vaccination programme is the view that the incidence of dental caries is falling in the West and that the disease can be adequately controlled using other techniques. As a result, vaccination is deemed unnecessary. However, it can be argued that not all Western countries are experiencing a decrease in caries rate and that, on a world-wide basis, a vast increase in caries may occur, especially in poor countries with little or no organized dental services. In this situation a safe and successful vaccine could be valuable in controlling the disease on a population basis. Also, prevention of disease in special high-risk groups, e.g. mentally or physically handicapped children, could be achieved.

It is not entirely clear how the vaccine produces its protective effect, although the following mechanisms have been suggested: (1) inhibition of the microbial colonization of enamel by secretory IgA; (2) interference with bacterial metabolism; and (3) enhancement of phagocytic activity in the gingival crevice area due to the opsonization of *Strep. mutans* with IgA or IgG antibodies. However, convincing proof that any of these mechanisms prevent the development of dental caries *in vivo* is lacking.

Fissure sealants

Fissure sealants prevent caries in pits and fissures by eliminating stagnation areas and blocking potential routes of infection leading deep within the tooth. The most commonly used method is the acid etch technique using an acrylic resin (Bowens resin) polymerized by either light or chemicals. When sealing is complete, the fissures are protected against invasion by oral microorganisms. If a fissure containing an early carious lesion is sealed, there is reasonable hope of a favourable result, since such lesions tend not to progress, probably because the source of microbial nutrition has been blocked. If a more extensive lesion is sealed, it is likely that if sufficient bacteria have already invaded the dentine, the lesion will not be arrested but extend into the pulp. The bacteria will obtain sufficient nutrients from the carious dentine.

Further reading

Carlsson, J. (1984) Regulation of sugar metabolism in relation to the feast and famine existence of plaque. In *Cariology Today*, Karger, Basel, pp. 205–211

Cohen, B., Peach, S. L. and Russell, R. R. B. (1983) Immunization against dental caries. In *Immunisation Against Bacterial Disease. Medical Microbiology*, Vol. 2, (eds C. S. F. Easmon and J. Jeljaszewicz), Academic Press, London, pp. 255–294

Edwardsson, S. (1986) Microorganisms associated with dental caries. In *Textbook of Cariology* (eds A. Thylstrup and O. Fejerskov), Munksgaard, Copenhagen, pp. 107–130

Krasse, B. (1985) *Caries Risk*, Quintessence, Chicago

Rugg-Gunn, A. J. (1983) Diet and dental caries. In *The Prevention of Dental Disease* (ed. J. J. Murray), Oxford Medical Publications, Oxford, pp. 3–82

Sims, W. (1985) *Streptococcus mutans* and vaccines for dental caries; a personal commentary and critique. *Community Dental Health*, **2**, 129–147

Chapter 5
Microbiology of periodontal diseases

Periodontal diseases occur in all parts of the world and few individuals live out their natural life span without becoming affected. However, in the majority of individuals, the common chronic inflammatory diseases which involve the gingival and periodontal tissues can be controlled, if not cured, using mechanical cleansing techniques and encouraging good oral hygiene. A small but significant number of patients experience rapidly progressive disease which requires assessment and treatment by periodontologists.

While there is no doubt that microorganisms play an important role in the aetiology of most forms of periodontal disease, there is dispute as to whether their involvement is of a specific or non-specific nature. At present, clinical and radiological examination are unable to diagnose active disease or to predict which patients are likely to experience severe progressive periodontitis. Tissue destruction can be recorded with certainty only in a retrospective manner. Therefore, much of the current research in periodontology is directed towards developing laboratory tests which will allow high-risk patients and active disease to be identified early and ensure that subsequent specialized treatment is effective. An area of current interest is AIDS-related periodontal disease which is described in Chapter 16.

Types of periodontal disease

The main types of periodontal disease which are believed to have a microbial aetiology are shown in Table 5.1. However, as detailed knowledge about the aetiology of some of these conditions, e.g. chronic periodontitis, becomes available, it is likely that periodontal infections with apparently the same clinical signs and symptoms will turn out to be due to a number of different host–parasite interactions.

Table 5.1 Classification of periodontal diseases with a microbial aetiology

Plaque associated
Gingivitis
  Acute: ulcerative (AUG)
          abscess
  Chronic: gingivitis
Periodontitis
  Acute: abscess
  Chronic: adult
          early onset
          : juvenile
          : prepubertal
          : rapidly progressive

Non-plaque associated
Acute herpetic stomatitis

Factors involved in periodontal diseases

The main aetiological factors of periodontal diseases are (1) the host tissues, (2) microorganisms in the form of subgingival plaque, and (3) the specific and non-specific host defence mechanisms.

Host tissues

The term periodontium refers to the anatomical structures involved in resisting forces applied to the teeth, especially the gingivae, periodontal ligament, cementum and alveolar bone. The part of the periodontium which supports the coronal portion of the root is known as the marginal periodontium (Figure 5.1). A detailed description of these tissues and the changes which occur with the onset of disease is outside the scope of this book and students should consult standard periodontology textbooks. While the dentogingival junction is a site of potential weakness, the host defences are able to function effectively as long as oral hygiene is satisfactory (Table 5.2). However, when plaque is allowed to accumulate undisturbed and close to the gingival margin, the host defences become stressed.

It is not known why some individuals are more susceptible to periodontal disease than others. One explanation may be related to genetic variations in the biochemical composition or anatomical arrangement of the periodontal tissues among individuals.

Factors involved in periodontal diseases 53

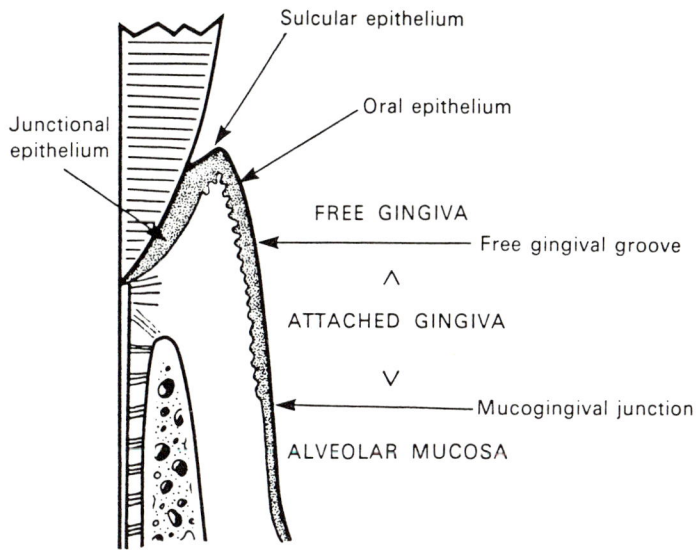

Figure 5.1 Healthy marginal periodontium. (From Jenkins, Allan and Collins, 1988, *Guide to Periodontics*, by permission)

Table 5.2 Specific and non-specific factors present in the gingival crevicular fluid

Specific	Non-specific
B and T lymphocytes	Polymorphonuclear and mononuclear phagocytes
Antibodies: IgG, IgM and IgA	Complement components Lysozyme Enzymes from host

However, very little is known about these factors.

Host defence factors

Both the specific and non-specific immune responses of the host to subgingival plaque are believed to play important roles in the development, progression and recovery from periodontal diseases. Although specific immune reactions may occur either in the gingival crevice/periodontal pocket, or within the periodontal tissues themselves, non-specific factors usually act in the former sites only.

One of the important components of the host response is the gingival crevicular fluid (GCF) which contains both specific and non-specific factors (Table 5.2). Although it is convenient to describe these factors separately, it is important to remember that *in vivo* they interact closely, e.g. antibodies can act as opsonins thereby potentiating phagocytic activity. The nature of the non-specific host factors are shown in Table 5.3.

Table 5.3 Non-specific defence mechanisms in the marginal periodontium

Host cells	Gingival crevicular fluid
Anatomical epithelial seal in the base of the gingival sulcus	Mechanical washing action; tendency to cleanse crevice
Rapid repair of junctional epithelium following injury	Antimicrobial factors, e.g. lysozyme
Shedding of surface cells with attached bacteria from sulcular epithelium into crevice	
Phagocytosis together with migration of polymorphonuclear and mononuclear phagocytes through the junctional epithelium	

Polymorphonuclear leucocytes
Although small numbers of polymorphonuclear leucocytes (PMNLs) are present in clinically healthy gingivae, there is a marked increase in numbers during the onset of gingivitis which persists during periodontitis. The PMNLs migrate from venules and enter the gingival sulcus through the junctional epithelial cells. The increased number found in periodontal disease is probably related to the chemotactic activity of complement, low molecular weight peptides and other plaque products.

When PMNLs come in contact with bacteria, phagocytosis begins. The polymorphs attach to and ingest the microbial cell, which becomes bound within the phagosome. The latter then fuses with a proteolytic and hydrolytic enzyme-rich lysosome to form a digestion vacuole, in which microbial killing occurs. In addition, other host-cell-derived killing agents may be involved, e.g. hydrogen peroxide and lactic acid. Although phagocytosis can occur in the absence of antibody, the presence of immunoglobulins and complement enhance the process.

The outcome of the interaction between PMNLs and plaque bacteria may result in either death of the microorganism, death of the leucocytes, or both. After neutrophil autolysis, lysosomal enzymes (e.g. collagenase, hyaluronidase, chondroitin sulphatase, elastase and acid hydrolases) are released around the surrounding host cells and may cause damage to the periodontal tissues.

Therefore, PMNLs may have both a protective and perhaps a damaging effect on host tissues. Patients who have compromised PMNL function due, for example, to agranulocytosis or cyclic neutropenia, usually develop aggressive periodontitis, which tends to support the concept that overall neutrophils are protective. Claims that patients with juvenile periodontitis and rapidly progressive periodontitis have defects in PMNL function should be accepted with caution, due to difficulties in the interpretation of assay results and failure to take into account the variation in values which can occur in the same individual from day to day.

Phagocytosis is probably important in preventing the microbial invasion of the gingival and periodontal connective tissues. However, some microbial antigens have been demonstrated within the periodontal tissues (e.g. *Treponema* spp, *A. actinomycetemcomitans* and *B. gingivalis*), and therefore it is likely that phagocytosis will take place within the host tissues and perhaps also at their interface with subgingival plaque.

Antibody
Specific IgM, IgG and IgA antibodies to subgingival plaque microorgainsims and some of their metabolites have been demonstrated in GCF. and serum. These antibodies are derived both from serum and from plasma cells in the gingival connective tissues (local antibody formation). The significance of an elevated titre of specific antibody to a periodontopathogen is uncertain at present. Various possibilities have been suggested, for example, that antibodies (1) are protective, (2) are involved in damaging hypersensitivity reactions to the host tissues, and (3) are non-specific and unrelated (i.e. epiphenomena). The presence of antibody suggests that the required T-helper and T-suppressor interactions with B lymphocytes occur satisfactorily in the periodontal tissues. In addition, all the cells required for a wide range of immune reactions have been demonstrated in the gingival tissues of chronic periodontitis patients and are known to possess antigen specificity for plaque bacteria. When stimulated either antibodies (from B lymphocytes) or lymphokines (from T lymphocytes) are produced.

Since antibodies and complement are present in the periodontal tissues, hypersensitivity reactions which could result in damage to host tissues may contribute to periodontal disease. While there is evidence that all four types of hypersensitivity may be involved under certain circumstances, their precise role in pathogenesis or in recovery remains uncertain (Table 5.4).

Microbial factors

There is no doubt that dental plaque is an essential component in the aetiology of the common forms of chronic gingivitis and periodontitis. The evidence for the role of plaque is briefly as follows:

1. Worldwide epidemiological studies have shown a strong positive association between plaque and the prevalence and severity of periodontal deseases.
2. Clinical studies in patients with a healthy periodontium have shown that, if oral hygiene is discontinued, the accumulation of dental plaque which occurs is paralleled by the onset of gingivitis. If plaque is then removed and normal oral hygiene reintroduced, the tissues are restored to health.
3. The topical application of certain antimicrobial agents both inhibit plaque formation and prevent the development of gingivitis.
4. Certain bacteria isolated from human dental plaque, e.g. *F. nucleatum, B. gingivalis* and *Eik. corrodens*, are periodontopathic in gnotobiotic animals.

The microbial composition of the gingival crevice area in health and disease is shown in Table 5.5. The flora associated with health consists mainly of streptococci and *Actinomyces* spp. This changes both quantitatively and qualitatively during the development of disease, and differences can be demonstrated in plaque samples from healthy and diseased sites, as well as between different types of periodontal diseases (for details, see below under specific diseases).

There has been research concerning the mechanisms by which plaque microorganisms may produce destruction of the periodontium. A wide range of microbial products which are potentially toxic to host tissues have been identified *in vitro* (Table 5.7). While there is evidence that some of these factors are produced *in vivo*, it is not clear what role they play in pathogenesis. Certainly, if all the potentially toxic products which have been described were constantly being released into the periodontal tissues, it would not be surprising that destructive inflammatory disease results. What is difficult to understand

Table 5.4 Possible hypersensitivity mechanisms in periodontitis

	Type 1 (anaphylactic)	Type 2 (cytotoxic)	Type 3 (immune complex)	Type 4 (cell-mediated)
Antibody-mediated (B lymphocyte)	+[a]	+[b]	+[b]	—
Main effector mechanisms	Histamine from mast cell degranulation. Vascular permeability increased	Components from complement activation produce increased vascular permeability, chemotaxis, phagocytosis and cell lysis	Antigen-antibody complexes deposited in tissues, especially blood vessel walls. Components of complement activation produce acute inflammation with possible necrosis	Lymphokines and killer T cells. Accumulation and activation of macrophages, inflammation and possibly bone resorption
Possible role in periodontitis	Uncertain	Uncertain	Uncertain	Uncertain

[a] IgE bound to mast cells
[b] Complement activation involved.

Table 5.5 Microorganisms associated with various forms of periodontal disease

Health/Disease	Predominant microorganisms	Comments
Health	Strep. sanguis Strep. oralis A. naeslundii A. viscosus Veillonella spp.	Mainly Gram-positive cocci with few spirochaetes or motile rods
Chronic gingivitis	Strep. sanguis Strep. milleri A. israelii A. naeslundii B. intermedius Capnocytophaga spp. F. nucleatum Veillonella spp.	About 65% of cells are Gram-positive with occasional spirochaetes and motile rods
Localized juvenile periodontitis (LJP)	A. actinomycetemcomitans Capnocytophaga spp. Eikenella corrodens	About 65% of bacteria are Gram-negative bacilli. Few spirochaetes or motile rods present
Rapidly progressive periodontitis (RPP)	A. actinomycetemcomitans B. gingivalis B. intermedius Corroding bacteria Spirochaetes	Possess elements of both LJP and chronic periodontitis
Chronic adult periodontitis	B. gingivalis B. intermedius F. nucleatum Strep. milleri Eubacterium spp. Eikenella corrodens Wolinella recta Spirochaetes	About 75% of cells are Gram-negative, (90% being strict anaerobes). Motile rods and spirochaetes are prominent. In refractory cases A. actinomycetemcomitans may be present

is why tissue destruction usually progresses so slowly, suggesting that very powerful host defence factors (about which little is known) are likely to be at work.

Although there is adequate evidence that plaque microorganisms play an important role in periodontal diseases, it is not certain if the destruction is due to a non-specific mixture of microorganisms, or if single species or specific complexes of a few different species are implicated. Overall it is likely that the specific and non-specific theories represent the two extremes of a complex series of host–parasite interactions, some of which will produce disease while others are conducive to health.

Dental calculus

Dental calculus is defined as calcified or calcifying deposits on the teeth and can form both above (supragingival) and below (subgingival) the gingival margin. It consists of 70–80% inorganic salts, mainly Ca and PO_4, the major crystalline component being three different types of hydroxyapatite, $Ca_{10}(PO_4)_6(OH)_2$. The organic portion consists mainly of proteins, carbohydrates and lipids. The structure and formation of dental calculus are shown in Figure 5.2.

Figure 5.2 The stages involved in the development of dental calculus. 1, enamel and calculus crystals in close contact, no pellicle; 2, intermicrobial matrix completely calcified with crystals in some bacteria; 3, some calcification of matrix, none in bacteria; 4, plaque with no evidence of calcification

Calculus is preceded by dental plaque which subsequently becomes calcified. The plaque matrix calcifies first, followed by the bacterial cells themselves. It follows that the outer surface of all calculus deposits is covered by a thin layer of viable dental plaque. There is evidence that the irritant effect of calculus on gingival and periodontal soft tissues is due more to the toxic potential of the viable plaque layer than to mechanical trauma from the hard irregular calculus deposits themselves. The rate of calculus formation varies considerably among individuals, with visible deposits being formed in some individuals 2–4 weeks after thorough scaling. The factors which trigger calcification of dental plaque are not clearly understood, although it seems likely that degenerating bacteria play an important role, perhaps by acting as seeding agents for mineralization.

Chronic gingivitis

Chronic gingivitis and chronic periodontitis are destructive inflammatory diseases. The lesions of the former are confined to the gingiva, while the tissue destruction of the latter involves both the connective tissue attachment of the tooth and loss of alveolar bone. It seems likely that, as more information collects about the aetiology of the factors involved in chronic periodontitis, several distinct types of disease will emerge which, although they may appear to be the same clinically, require different forms of treatment.

Gingivitis is common both in adults and children, although early periodontitis is rarely seen before late adolescence. Information concerning the natural history of periodontal diseases is incomplete. Therefore, although it is assumed that chronic periodontitis is preceded by chronic gingivitis, this is based mainly on clinical experience and cross-sectional surveys, rather than on good experimental data which describes the transitional stage from gingival to periodontal destruction.

The main stages in the development of chronic gingivitis and periodontitis are shown diagrammatically in Figure 5.3. However, it must be remembered that much of the histological information has been obtained from animal experiments and care must be exercised in directly extrapolating the results to spontaneous disease in humans.

Clinical presentation

The main clinical feature of chronic gingivitis is red, swollen gingiva with rounded edges. While bleeding gums and halitosis are common, pain, discomfort and unpleasant taste are not.

Within the first few days the main histological changes in

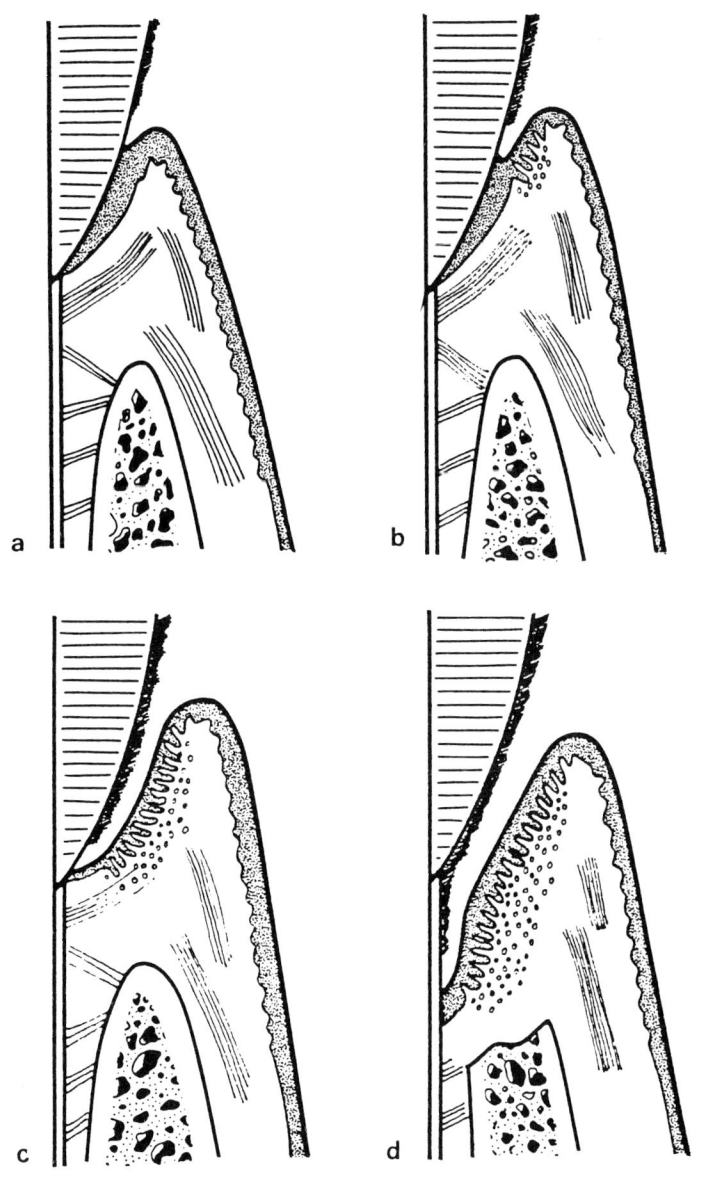

Figure 5.3 The development of chronic gingivitis and periodontitis. (a) A healthy gingival sulcus with early supragingival plaque formation; (b) established chronic gingivitis with minor inflammatory enlargement; (c) long-standing chronic gingivitis with subgingival plaque extension in a gingival pocket; (d) chronic periodontitis with destruction of periodontal membrane and alveolar bone and apical migration of epithelial attachment. (After Jenkins, Allan and Collins, 1988, *Guide to Periodontis*, by permission)

gingivitis include vasculitis, perivascular collagen destruction and an increase in crevicular fluid and polymorphonuclear leucocytes in the junctional epithelium and crevice. At this stage no clinical change is evident. By the 7th day a dense infiltration of lymphocytes (mainly T cells) appears, but by 10–21 days B lymphocytes and plasma cells predominate, the cellular infiltrate being chronic in nature. Further collagen destruction occurs and the inflammatory process slowly spreads to involve more tissue. Clinically recognizable chronic gingivitis is now present, with shallow gingival pockets which bleed on probing. If plaque is not removed and oral hygiene improved, this condition may persist for many years without extending into the deeper periodontal tissues.

Microbiology

The changes which take place in the microflora of the gingival crevice region with the onset of chronic gingivitis are shown in Table 5.5. There is general agreement that gingivitis is related to the prolonged exposure of host tissues to a non-specific mixture of gingival plaque organisms. The precise microbial composition of the plaque does not appear to be important and specific pathogens have not been identified. However, the onset of inflammation tends to be related to an increase in the numbers of anaerobic Gram-negative rods. The biomass of plaque appears to be related to the development of gingivitis, with the penetration of the crevicular epithelium by extracellular microbial products, leading to gingival inflammation.

Treatment

Treatment involves the thorough removal of plaque and calculus deposits, and the introduction of good oral hygiene.

The transitional stage

It has been estimated that chronic gingivitis may be present for 5–11 years before progressing to periodontitis. A variety of suggestions have been made to explain this transition, although there is little evidence to support any of them. The suggestions include: (1) selective overgrowth of one or more plaque species due to some impairment of the host defences; (2) infection and proliferation of a new pathogen in the gingival area; and (3) activation of tissue-destructive immune processes which may be related to (1) above or other unknown factors.

Chronic periodontitis

While a detailed description of the tissue changes which occur is outside the scope of this book, the main processes which produce loss of attachment and pocket formation (Figure 5.3) are as follows: (1) the apical spread of subgingival plaque causes the junctional epithelium to separate from the tooth surface, thus becoming 'pocket' epithelium; (2) inflammatory reactions in the tissues below the pocket epithelium are associated with destruction of the gingival connective tissue, periodontal membrane and alveolar bone; (3) apical proliferation of the junctional epithelium results in migration of the epithelial attachment.

Overall, chronic periodontitis is characterized by destruction of the connective tissues attached to the teeth and the production of progressively deeper pockets. However, the rate of destruction is not constant and periods of quiescence alternate with bouts of bone resorption. Currently it is thought that a number of patterns of disease activity can occur, ranging from slowly progressive destruction to brief bursts of activity which may vary in intensity and duration in different sites in the same mouth.

Unfortunately, the diagnosis of these disease patterns and therefore the periods when tissue destruction is actually taking place in any particular patient can only be made retrospectively, using current clinical and radiological criteria.

Clinical description

A wide range of clinical features may be associated with chronic periodontitis and include gingival inflammation and bleeding pockets of more than 4 mm, tooth mobility and migration, alveolar bone loss, gingival recession, halitosis and offensive taste. Usually patients experience little pain.

Microbiology

A summary of the microorganisms associated with chronic periodontitis is shown in Table 5.5. Microbiological studies of plaque from patients with chronic periodontitis have involved one or more of the following techniques:

1. Dark-field microscopy of subgingival plaque to estimate the different morphological bacterial types present, especially spirochaetes and other motile bacteria.
2. Cultural studies using screening methods for the presence of selected periodontopathic microorganisms.
3. In-depth microbiological studies using conventional culture techniques to isolate, identify and enumerate all microbial species in plaque.
4. Studies using molecular biology techniques, e.g. specific DNA probes.

It must be understood, however, that one of the major problems in any study which attempts to prove that specific microorganisms cause chronic periodontitis, is the absence of clinical, radiological or laboratory tests which can accurately identify periods when destructive disease is taking place. This means that it is difficult to be certain that any given sample has been collected from an active site. Since most sites in the majority of mouths will be inactive at any given time, the chance of success must be low. Another problem is related to sample collection. It seems likely that periodontopathic bacteria will be present in the depths of a periodontal pocket close to the site where destruction is occurring. However, it is very difficult to ensure that the samples collected in fact come from the base of the pocket. In addition, the anatomy of periodontal pockets are very variable and it is likely that gross contamination of the sample with overlying plaque will occur during collection. While keeping these problems in mind, let us now consider the evidence for specific bacteria being involved in the initiation and progression of chronic periodontitis.

Spirochaetes
Since the majority of oral spirochaetes cannot be isolated, identified and enumerated by culture methods, dark-field microscopical methods have been used to study them. The results of these investigations have shown that significantly lower numbers of spirochaetes are present in healthy, compared with diseased, periodontal sites. For sometime it was believed that a high percentage of spirochaetes in a subgingival sample strongly suggested a site undergoing, or about to experience, active destructive disease. Over the years it has become clear that the percentage of spirochaetes in a sample cannot be used reliably to identify or predict active periodontitis. As a result, the evidence for spirochaete specificity is conflicting and confused.

Lack of correlation with disease does not of course prove that spirochaetes are not involved in the pathogenesis of chronic periodontitis. It is possible that one or more *Treponema* spp. are involved aetiologically but that the majority of species are not. Microscopical counts of all spriochaetes as performed at present would fail to identify such a relationship. However, the role of specific spirochaetes in periodontal diseases cannot be proven or refuted until accurate laboratory techniques have been developed.

Black pigmented Bacteroides *species*

At least six different *Bacteroides* spp. which produce black pigmented colonies on blood agar have been isolated from the oral cavity (Table 5.6). However, most studies have investigated the roles of only *B. gingivalis* and to a lesser extent *B. intermedius* in periodontal disease. One of the main reasons why these species have been used as screening organisms is that the distinctive pigmented colonies which they produce can be easily counted and simple identification tests can be subsequently performed.

Table 5.6 Classification of black pigmented *Bacteroides* species found in the human mouth

Non-fermenter	Weak fermenter	Strong fermenter
B. gingivalis	B. intermedius	B. melaninogenicus
B. endodontalis		B. denticola
		B. loescheii

The evidence for specificity of *B. gingivalis* in chronic periodontitis depends mainly on the following: (1) clinical association studies; (2) the production of a wide range of factors *in vitro* which can impair the host defences and damage components of the periodontium; and (3) infections in experimental animals which have produced both soft tissue destruction and bone resorption.

A list of the pathogenic factors produced by *B. gingivalis in vitro* is shown in Table 5.7. However, while it seems likely that *B. gingivalis* may be implicated in tissue destruction in a proportion of patients with advanced chronic periodontitis, there is insufficient evidence to extrapolate these results to the disease in general.

Table 5.7 Potentially pathogenic factors produced by *Bacteroides gingivalis* **and** *Actinobacillus actinomycetemcomitans*

	B. gingivalis	A. actinomycetemcomitans
Degradation of IgA and IgG	+	−
Collagenase	+	+
Hyaluronidase	+	−
Endotoxin	+	+
Fibroblast growth inhibitors	+	+
Leucotoxin	−	+

Capnocytophaga *and corroding bacteria*

Capnocytophaga spp. are members of the commensal oral flora and it is doubtful if they should now be regarded as prime pathogens in periodontal or oral infections. Various corroding bacteria, e.g. *Wolinella* spp. and *Eik. corrodens*, have been associated with a number of forms of periodontal disease. However, their precise role is uncertain and further investigations are required to clarify their importance.

Detailed culture studies

Relatively few studies of this type have been performed. The results generally have not supported a specific microbial aetiology for periodontal diseases. For example, one study reported over 56 microbial species which were more commonly isolated, and in higher numbers, from diseased compared with healthy sites. The list included bacteroides, spirochaetes, fusobacteria, lactobacilli, eubacteria and anaerobic streptococci. More recently *Wolinella recta* has been related to sites where active disease was present. This confusing data can be interpreted as good evidence for a non-specific aetiology of chronic periodontitis. Another explanation is that much of the confusion is due to the problems associated with the accurate diagnosis of active disease, together with sampling and laboratory difficulties as discussed earlier. If these problems could be solved, then the extent of microbial specificity in chronic periodontitis could become clear.

'Refractory' chronic periodontitis

There is evidence, not fully corroborated, that *A. actinomycetemcomitans* can be regularly isolated from subgingival plaque samples collected from adult patients who have chronic periodontitis which cannot be controlled by conventional mechanical therapy. Usually, if a combination of mechanical therapy and tetracycline is used to treat such patients, the progress of the disease is stopped and clinical improvement occurs. However, on the information available it is not certain if the presence of this organism in these refractory lesions is cause or effect.

Localized juvenile periodontitis

Clinical appearance

The onset of localized juvenile periodontitis (LJP) occurs in about 0.1% of otherwise healthy adolescents; it occurs more commonly in females and in individuals of African or Asian origin. The disease is characterized by a rapid destruction of the periodontal tissues. Initially the first permanent molars and permanent incisors are affected but, in time (usually in the early twenties), other teeth may become involved, producing the appearance of generalized alveolar bone loss. In some cases about 50% of the supporting alveolar bone is affected and teeth may be lost. Usually oral hygiene is satisfactory and, in contrast to chronic periodontitis, little plaque or calculus is present in periodontal pockets.

Aetiology

Arguably the best evidence for microbial specificity is found in LJP. There is some evidence that the disease is inherited, although no convincing defect either in the host tissues or defence mechanisms has been defined. Another explanation for the features of the disease is that it is caused by infection with a specific microorganism *Actinobacillus actinomycetemcomitans*, either alone or synergistically with *Capnocytophaga* spp. and *Bacteroides* spp. (especially *B. gingivalis*). There is doubt about the significance of *Capnocytophaga* spp. in LJP since it is now clear that these capnophilic rods are common members of the oral microflora.

Microbiology

The majority of subgingival plaque samples obtained from patients with LJP are different both qualitatively and quantitatively from the majority of patients with chronic periodontitis (see Table 5.5). The evidence for the specific involvement of *A. actinomycetemcomitans* in LJP includes the following:

1. A close correlation between the presence of *A. actinomycetemcomitans* in the subgingival plaque of sites affected by localized juvenile periodontitis.
2. Evidence of invasion by *A. actinomycetemcomitans* into the soft tissues of the periodontium.
3. Immunological studies which have demonstrated high levels of antibody to *A. actinomycetemcomitans* which tend to fall after successful treatment.
4. *In vitro* demonstration of a wide range of potentially pathogenic products ideally suited to a periodontopathic microorganism. However, there is evidence that not all strains are equally leucotoxic, which raises the possibility that, for accurate diagnosis, isolation and identification may require to be combined with tests for toxigenicity to polymorphonuclear leucocytes.
5. Successful treatment with tetracycline is associated with elimination of the organism from diseased sites.

A. actinomycetemcomitans is a rarely reported but recognized pathogen in medical microbiology, being involved in cases of actinomycosis (see Chapter 6), abscesses of the abdomen, hand and brain, as well as heart lesions in cases of infective endocarditis.

Treatment

Mechanical therapy remains the main way of treating this form of periodontitis. However, there is evidence that in many cases adjunct therapy consisting of tetracycline in a dosage of 250 mg three times a day for four weeks produces resolution of the disease and may reduce the risk of reactivation at a later date.

Prepubertal periodontitis

Prepubertal periodontitis occurs both in a localized and generalized form and can affect both the primary and secondary dentition. The disease is rare and, while the localized form responds well to mechanical débridement with penicillin therapy (250mg q.i.d.), the generalized form tends to be difficult to control.

Neither *A. actinomycetemcomitans* nor *B. gingivalis* appear to be associated with this form of periodontal disease, although relatively few cases have been studied in detail. Culture studies suggest that *B. intermedius* and *Selenomonas sputigena* are commonly isolated.

Rapidly progressive periodontitis

Rapidly progressive periodontitis (RPP) is most common in young adults (20–35 years) and may be a late stage of LJP. The disease is characterized by phases of active destruction with marked gingival inflammation and rapid bone loss followed by periods of remission. Some patients respond well to mechanical therapy and antimicrobial therapy, while others do not.

Quantitatively and qualitatively, subgingival plaque from sites affected by this disease is more comparable to that described for chronic adult periodontitis (see Table 5.5) than for juvenile periodontitis.

Clinical implications of microbiological tests

It has been suggested that microbiological tests could be used in the management of periodontal disease for the following purposes: (1) to identify sites of active tissue destruction; (2) to help in planning treatment for both new and refractory patients; and (3) to monitor the effects of treatment and decide when recall is necessary. However, these aims can only be fulfilled if the causative pathogens are definitively known, and if suitable laboratory tests are available to identify and quantitate their presence in samples collected accurately from the site of disease (i.e. probably the base of the pocket). It is clear that this stage has not yet been reached and much doubt exists about almost all of the above stages in diagnosis. Sampling for the presence of *A. actinomycetemcomitans* in juvenile periodontitis, and perhaps refractory chronic periodontitis patients, is the only microbiological test which is likely to contribute to the treatment of chronic periodontal diseases at present. A positive test would suggest that tetracycline therapy should be used as an adjunct to root instrumentation.

The use of spirochaetes or *B. gingivalis* as marker organisms for monitoring treatment may be useful in certain situations.

Treatment of chronic periodontitis and juvenile periodontitis

The standard method of treating chronic periodontitis remains mechanical therapy, with or without periodontal surgery. However, adjunct therapy with antibacterial agents has gained in popularity. The agents used include: (1) pocket irrigation with chlorhexidine or metronidazole; (2) slow-release devices, containing either tetracycline, metronidazole or chlorhexidine, which are introduced into periodontal pockets; and (3) the systemic use of tetracycline or metronidazole. It is difficult to give definite advice about the indications for the use of these agents since the literature on this topic is conflicting and confusing. Probably the clearest indication for antimicrobial therapy is where *A. actinomycetemcomitans* has been isolated from a suspected LJP, RPP or an adult with refractory chronic periodontitis, when tetracycline (250 mg t.i.d. for 4 weeks) may be used to some advantage.

Acute ulcerative gingivitis

Clinical presentation

Acute ulcerative gingivitis (AUG) is characterized by acutely inflamed gingivae (red, shiny and bleed easily) with irregularly shaped ulcers which initially appear on the tips of the interdental papillae. If untreated, the ulcers may enlarge and spread to involve the marginal and, in extreme cases, the attached gingivae. The lesions are extremely painful and are covered by a pseudomembrane or slough which tends to break up as it is wiped from the surface. The slough consists of leucocytes, erythrocytes, fibrin, necrotic tissue debris and microorganisms. In most cases of AUG the patient's breath has a very unpleasant odour; patients also report an unpleasant metallic taste. Usually patients experience little systemic upset and, while there may be mild submental or submandibular lymphadenitis, involvement of the cervical lymph nodes occurs only in severe cases. Generalized fever or malaise is very uncommon.

If AUG is inadequately treated, the rate of tissue destruction is reduced but the disease enters a chronic phase in which pronounced loss of supporting tissues can occur over a period of time.

Aetiology

The main factors which predispose to the onset of AUG, are

shown in Table 5.8. Many of these factors are inter-related, e.g. heavy smokers tend also to have poor oral hygiene and suffer from stress. In addition, it has been suggested that cytomegalovirus may be aetiologically involved, possibly by depressing the cell-mediated response of the host.

Table 5.8 Factors predisposing to acute ulcerative gingivitis

Inadequate oral hygiene with pre-existing gingivitis
Heavy smoking
Emotional and psychic stress
Secondary to primary herpetic gingivostomatitis
Acquired immunosuppression:
   recent infection with measles virus
   previous infection with human immunodeficiency virus (HIV)
Severe malnutrition

Microbiology

The characteristic microscopical appearance of ulcer debris collected from a case of AUG is shown diagrammatically in Figure 5.4. As a result, a strong association has developed between this fuso–spirochaetal complex and AUG. The main evidence for specificity includes: (1) microscopical association studies; (2) the ability of the complex to cause tissue destruction in other body sites, e.g. tonsil (Vincent's angina); (3) recent animal studies; (4) the fact that treatment with metronidazole leads to a rapid elimination of the fuso–spirochaetal complex with associated obvious clincial improvement; and (5) invasion

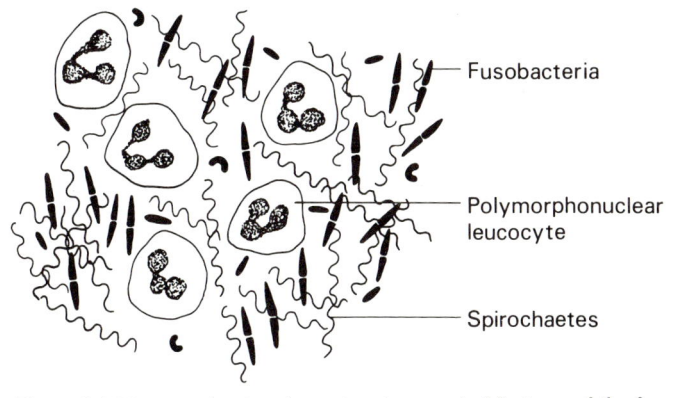

Figure 5.4 Diagram showing the main microscopical features of the fuso–spirochaetal complex associated with acute ulcerative gingivitis

of the gingival soft tissues by both spirochaetes and fusiform bacilli.

Surprisingly, there have been very few culture studies of samples from AUG patients and results suggest that medium-sized spirochaetes account for about 32% of the total microflora, *B. intermedius* for 24%, a spiral-like *Selenomonas* spp. for 6%, and fusobacteria for only 3%. In addition, a wide range of bacteria, e.g. streptococci, veillonella and *Actinomyces* spp., have been isolated in smaller numbers. A possible way in which the host defences, certain predisposing factors and the subgingival microflora may interact to produce tissue necrosis is shown in Figure 5.5.

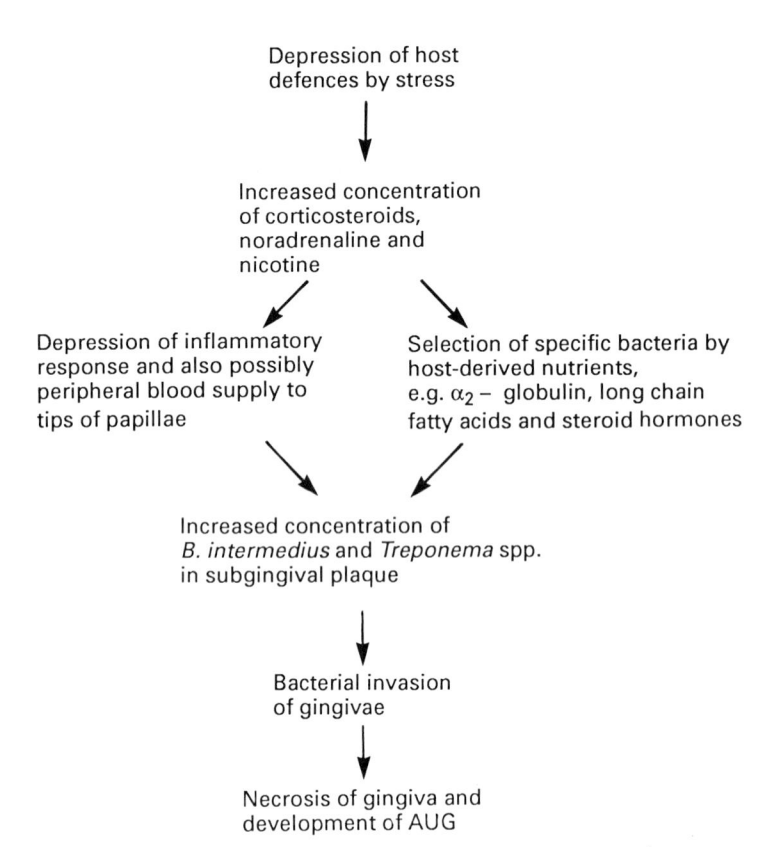

Depression of host defences by stress
↓
Increased concentration of corticosteroids, noradrenaline and nicotine
↙   ↘
Depression of inflammatory response and also possibly peripheral blood supply to tips of papillae

Selection of specific bacteria by host-derived nutrients, e.g. α_2 – globulin, long chain fatty acids and steroid hormones
↘   ↙
Increased concentration of *B. intermedius* and *Treponema* spp. in subgingival plaque
↓
Bacterial invasion of gingivae
↓
Necrosis of gingiva and development of AUG

Figure 5.5 Possible host–parasite interactions in the development of acute ulcerative gingivitis

Diagnosis

The diagnosis is usually made on clinical grounds but can be confirmed by microscopical examination of a smear prepared from the surface of the ulcerated lesion. Typically, a Gram-stained film shows a predominance of leucocytes, medium-sized spirochaetes and fusobacteria, and the presence of all three components must be present before a microbiological diagnosis can be confidently given. The differential clinical diagnosis includes primary herpetic stomatitis, gonococcal gingivitis, benign mucous membrane pemphigoid, desquamative gingivitis and some forms of leukaemia.

Treatment

While treatment varies with the severity of the disease, local therapy is extremely important in all cases and débridement should be initiated using ultrasonic scaling. Oral hygiene advice should be given and mouthwashes, e.g. chlorhexidine, may be a useful adjunct in some cases. Metronidazole is the drug of choice, in a dosage of 200 mg three times a day for 3 days. To ensure a rapid and long lasting cure it is important that both local and systemic treatment are given concurrently.

Noma

In underdeveloped countries an extremely severe form of AUG called noma or cancrum oris is found in children. Typically the child is less than 10 years of age, is malnourished (especially with regard to protein) and has a recent history of viral infection, e.g. measles. A possible important result of this combination of malnutrition and viral infection is depression of the child's specific immune mechanisms. The initial necrotic lesion spreads locally from the gingivae into the cheek and may extend on to the surface of the face, causing extensive tissue loss and severe disfigurement. This form of AUG is extremely rare in developed countries.

Further reading

Grant, D. A., Stern, I. B. and Listgarten, M. A. (1988) *Periodontics*, 6th edn, Mosby, St. Louis, pp. 252–292
Johnson, B. D. and Engel, D. (1986) Acute necrotizing ulcerative gingivitis. *Journal of Periodontology*, **57**, 141–150
Moore, W. E. C. (1987) Microbiology of periodontal disease. *Journal of*

Periodontal Research, **22**, 235–341

Slots, J. and Listgarten, M. A. (1988) *Bacteroides gingivalis, Bacteroides intermedius* and *Actinobacillus actinomycetemcomitans* in human periodontal disease. *Journal of Clinical Periodontology*, **15**, 83–93

Van Winkelhoff, A. J., Van Steenbergen, T. J. M. and de Graaff, J. (1988) The role of black-pigmented *Bacteroides* in human oral infections. *Journal of Clinical Periodontology*, **15**, 145–155

Chapter 6
Dentoalveolar infections

Pus-producing or pyogenic infections which are associated with a tooth and surrounding supporting structures, such as the periodontal membrane, cementum and alveolar bone, have been described using a number of terms such as periapical abscess, apical abscess, chronic periapical dental infection, dental pyogenic infection, periapical periodontitis, and dentoalveolar abscess. In this chapter the last term will be used. The clinical presentation is variable and is related to the interaction of a number of factors such as the virulence of the causative microorganisms, the state of the local and systemic defence mechanisms of the host and a number of anatomical features. Depending on how these factors interact, the resulting infection may present as an abscess localized to the tooth which initiated the infection, as a diffuse cellulitis which spreads along fascial planes, or a mixture of both. The primary source of microorganisms in these infections is either from the apex of a necrotic tooth or from periodontal pockets. Usually the source of infection in the periapical region is from dental caries via the pulp chamber and root canal.

Dentoalveolar abscess

Aetiology

The usual stages in the development and spread of a dentoalveolar abscess are shown in Figure 6.1. Microorganisms present in the initial carious lesion extend into dentine and spread to the pulp via the dentinal tubules. The pulp responds to infection in a number of ways, varying from rapid acute inflammation involving the whole pulp which quickly becomes necrosed, to the development of a chronic localized abscess with most of the pulp remaining viable. It is not clear why these different responses occur but it is probably due to a combination of a number of factors, such as variation in the toxicity of different bacterial species, as well as in the specific and non-specific defence mechanisms of the host. Microorganisms may enter the pulp by routes other than through a carious lesion, e.g. invasion may occur as a result of tooth fracture or as a result of a traumatic exposure during dental treatment. Pulp death may occur as a result of trauma without fracture or by chemicals used incorrectly during restorative treatment. In these examples the dead pulp tissue is sterile but may subsequently become infected via the gingival lymphatics or as a sequel to a bacteraemia caused, for example, by tooth extraction at a different site.

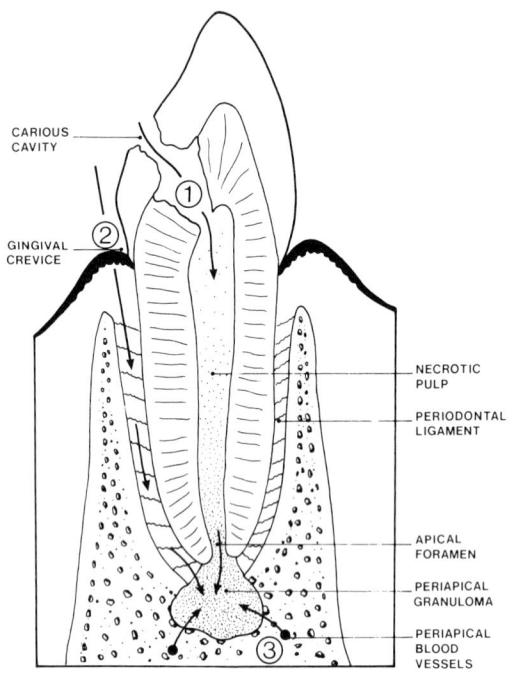

Figure 6.1 The pathways by which microorganisms may invade the pulp and periapical tissues. 1, from the apical foramen; 2, via the periodontal ligament; 3, via the blood stream. (Courtesy of Dr M. A. O. Lewis)

Infection and death of the pulp results in the pulp chamber and root canal becoming colonized by microorganisms. These microorganisms have a potential to produce a wide range of irritant substances, for example, enzymes, acids and toxins. Bacteria and their irritant products may leak from the root canal into the periapical tissues and stimulate an acute inflammatory response. As in the case of the dental pulp, the periapical tissues may react to infection in a number of ways depending on the virulence of the microorganisms involved and the efficiency of the hosts specific and non-specific defence mechanisms (see Chapter 3). However, the precise reasons why an acute abscess develops during periapical infection in some individuals but not others remain unclear.

Spread of infection and sequelae

Pus, once formed, may remain localized at the root apex with the formation of a chronic abscess which can develop into a focal osteomyelitis, or may spread by various routes as shown in Figure 6.2. Pus may spread through the cortical bone of the jaws into the superficial soft tissues or, more rarely, into the

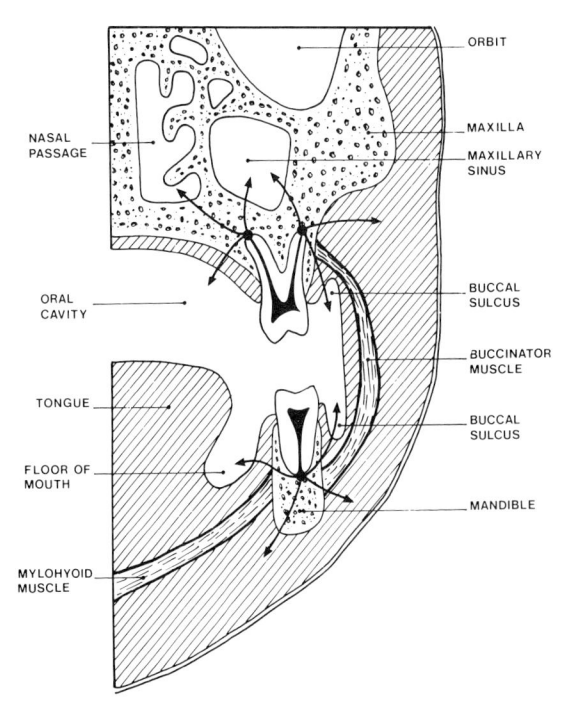

Figure 6.2 Pathways (arrowed) by which pus may track from an acute dentoalveolar abscess (coronal section of left side, at plane of first molar tooth). (Courtesy of Dr M. A. O. Lewis)

adjacent fascial spaces. Once in the soft tissues the pus may (1) localize as a soft tissue abscess, (2) spread through the overlying oral mucosa or skin producing a sinus linking the main abscess cavity with the mouth or skin, or (3) extend through the soft tissue to produce a cellulitis. When dentoalveolar infection spreads deeply into soft tissues it tends to follow the path of least resistance, i.e. through connective tissue and along fascial planes. Extension of infection into deep fascial spaces is determined by the relation of the original abscess to the anatomy of the adjacent tissues. Students interested in the topic should consult a textbook on oral surgery. Infection via fascial planes often spreads rapidly and for some distance from the original abscess site and occasionally may cause severe respiratory distress, due to occlusion of the airway by oedema. Finally, pus may spread into the deeper medullary spaces of alveolar bone, producing a spreading osteomyelitis.

Clinical presentation

The presenting features of acute dental alveolar abscesses are very variable depending on (1) the site of infection, (2) the degree and mode of spread, (3) the virulence of the causative microorganisms, and (4) the efficiency of the host defence mechanisms. Usually a mixture of some or all of the following features are present: a non-viable tooth with or without a carious lesion; a large restoration; evidence of trauma; swelling; pain; redness; trismus; local lymph node enlargement; sinus formation; raised temperature and malaise due to toxaemia. A photograph of a typical dentoalveolar abscess is shown in Figure 6.3.

Microorganisms involved

A microbiological investigation on a non-contaminated sample of pus from a dentoalveolar infection may yield a single isolate, a mixture of two to three different bacterial species, or a complex mixture of microorganisms involving perhaps eight different species. A single isolate is unusual and a mixture of 3–4 different species much more common. Many earlier reports dealing with the microbiology of dentoalveolar abscesses have undoubtedly underestimated the incidence of strictly anaerobic bacteria, probably due to a combination of inadequate sampling and transportation to the laboratory, together with poor laboratory technique. Well-controlled studies have shown that strict

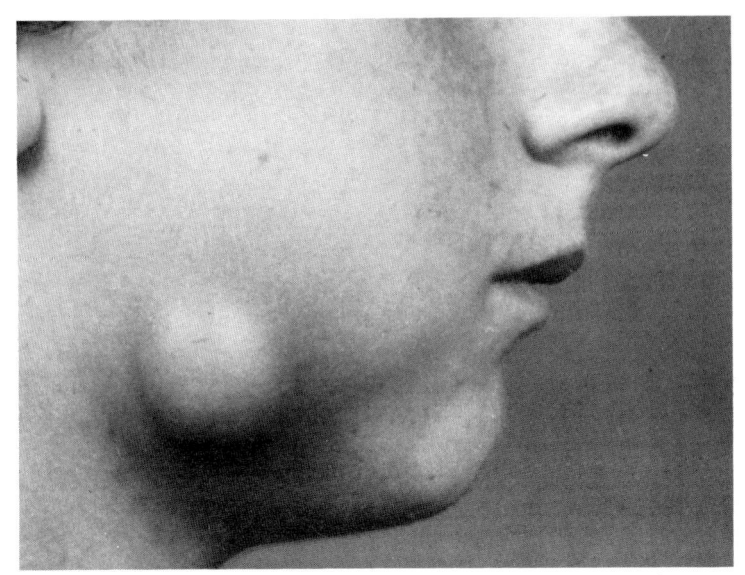

Figure 6.3 Localized swelling of right cheek due to an acute dentoalveolar abscess affecting the lower right first molar

anaerobes are usually the predominant organisms and that '*Streptococcus viridans*' is less common than one would assume from many reports. While almost every member of the oral flora has been isolated from dentoalveolar infection at one time or another, there is little doubt that some bacterial species are more common than others. The common species isolated from non-specific abscesses are shown in Table 6.1 and it is clear that strict anaerobes predominate (especially *Bacteroides* spp. and anaerobic streptococci), although facultative bacteria, e.g. *Strep. milleri*, are also found both in pure and mixed culture.

There is little information concerning the *in vivo* pathogenic potential of the bacteria commonly isolated from dental abscesses and therefore, until proven otherwise, all isolates should be regarded as equally important. Of course this is unlikely to be true, and there is good evidence that some strictly anaerobic bacteria, especially *B. gingivalis, B. endodontalis* and *Fusobacterium* spp. are more likely to cause severe infection than other bacterial species. There are a number of examples in medical microbiology where mixtures of strict anaerobes and aerobic or facultative bacteria are more pathogenic then either species alone. Such synergistic microbial interactions are likely to be important in the severity of dentoalveolar abscesses.

Table 6.1 Identity of 166 bacterial strains isolated from 50 acute dentoalveolar abscesses

Facultative anaerobes	Number of isolates	Strict anaerobes	Number of isolates
Streptococcus milleri	25	Peptostreptococcus spp.	14
Streptococcus mitior	3	Peptococcus spp.	32
Streptococcus sanguis	3	Streptococcus intermedius	3
Streptococcus mutans	1	Streptococcus constellatus	1
Lactobacillus fermentum	2	Propionibacterium acnes	1
Lactobacillus salivarius	1	Eubacterium lentum	1
Actinomyces spp.	3	Veillonella parvula	3
Arachnia propionica	1	Bacteroides oralis	20
Haemophilus parainfluenzae	2	Bacteroides gingivalis	14
Capnocytophaga ochracea	1	Bacteroides melaninogenicus	12
Eikenella corrodens	1	Bacteroides intermedius	5
		Non-pigmented bacteroides	10
		Fusobacterium nucleatum	6
		Fusobacterium mortiferum	1
Total	43	Total	123

Collection and transportation of samples

The ideal sample of pus for microbiological assessment is a non-contaminated fluid sample which is transported directly to the microbiology laboratory. Details of the methods of sampling dentoalveolar infections and transporting the samples to the laboratory are described in Chapter 13. Wherever possible, pus should be collected by needle aspiration, or in a sterile container after external incision. If swabs must be used then a strict aseptic collection technique is required, and if the pus sample is thought to have been contaminated with saliva or dental plaque during collection, this information must be recorded on the request form.

Treatment

Treatment of a dentoalveolar abscess is based on drainage of pus, removal of the source of infection and treatment with antibiotics. The specific treatment in any given individual will vary, depending on the clinical presentation of the lesion and the general health of the patient. If the abscess is localized within bone in the periapical area, drainage is accomplished by extraction of the tooth. If the tooth is to be retained, drainage should be established through the pulp chamber. Antibiotic therapy is probably not required for the majority of localized abscesses, although it may be necessary when root canal therapy is planned. If the abscess has spread to the superficial soft tissues of the mouth or face, antibiotic therapy should be included in patient management. The results of a number of microbiological studies suggest that a conventional 5-day course of penicillin V (250 mg t.d.s) or short-course, high-dose amoxycillin (two 3 g sachets of amoxycillin, the second taken 8 hours after the first) are suitable in the management of most abscesses. However, when penicillin therapy is not appropriate, e.g. in hypersensitive patients, erythromycin (250 mg 6-hourly for 5 days) or metronidazole (200 mg 8-hourly for 5 days) should be used. The use of clindamycin has been advocated for abscesses but its use should be limited to specific, severe infections, due to the well-defined relationship of clindamycin therapy with antibiotic-associated colitis.

The sequelae which can occur if a dentoalveolar abscess is untreated are shown in Table 6.2. Some of these conditions will be discussed later in this chapter.

Table 6.2 Sequelae of untreated localized dentoalveolar abscesses

Spread locally into alveolar bone producing an acute or chronic osteitis
Spread through bone or between fascial planes to give:
  localized acute or chronic soft tissue abscess
  acute spreading infections (cellulitis and fascitis)
Chronic abscess with acute exacerbations with discharging sinus on to oral mucosa or skin
Systemic spread by erosion into a blood vessel with resulting bacteraemia or septicaemia
Periapical granuloma and cyst formation

Ludwig's angina

Aetiology

In about 90% of cases with Ludwig's angina the primary source of infection is of dental or postextraction origin. Other sources of infection are submandibular sialadenitis, infected mandibular fracture, oral soft tissue laceration, and puncture wounds of the floor of the mouth.

Clinical presentation

The infection is characterized by a bilateral infection of the sublingual and submandibular spaces which raises the floor of the mouth and tongue and distends the tissues at the front of the neck. The areas of swelling have a characteristic board-like consistency and can barely be indented by the finger. There is severe systemic upset with fever, and complications include airway obstruction due either to oedema of the glottis or to a swollen tongue blocking the nasopharynx. Attempts at surgical drainage may yield little pus, and infection may spread to involve the masticator and pharyngeal spaces.

Source and type of microorganisms

Due to the severity of this condition, samples for microbiology assessment should be obtained from all cases. In the past β-haemolytic streptococci were isolated from patients with Ludwig's angina. At present, oral commensal bacteria are more commonly isolated, especially *Bacteroides* spp., *Fusobacterium* spp. and anaerobic streptococci. Pus from these infections usually contains a mixture of microorganisms, and occasionally coliforms such as *Escherichia coli* and *Klebsiella* species have been reported.

Treatment

It is important that the diagnosis is made as soon as possible and that the condition is not confused clinically with submandibular gland infection. If necessary, steps should be taken to ensure that the patients airway remains open and that his fluid balance is maintained. A sample of pus should be collected and high dosage empirical antibiotic therapy (usually penicillin, with or without metronidazole) started immediately. The antibiotic may be changed once the bacteriological report becomes available. Surgical drainage should be carried out as soon as possible and the primary source of infection, for example a non-vital tooth, eliminated.

Periodontal abscess

Aetiology

Knowledge about the aetiology and pathogenesis of periodontal abscess is incomplete. It is thought that this type of abscess forms by occlusion or trauma to the orifice of a periodontal pocket which results in the extension of infection from the pocket into the supporting tissues. Normally the abscess remains localized in the periodontal tissues. The subsequent development of the lesion depends on the interaction of the following factors: (1) the virulence, type and number of the causative organisms; (2) the existing health of the patient's periodontal tissues; and (3) the efficiency of the specific and non-specific defence mechanisms of the host. Factors such as the impaction of food (e.g. a fish bone) or of a detached toothbrush bristle have been related to the development of the infection. Other aetiological factors include recent scaling or compression of the pocket wall by orthodontic tooth movement or by unusual occlusal forces.

Spread of infection

Pus from this type of abscess usually drains along the root surface to the orifice of the periodontal pocket, although in the

84 Dentoalveolar infections

case of a deep pocket it may extend through the alveolar bone to drain via a sinus which opens on to the attached gingiva. Because drainage usually occurs, infection tends to remain localized and extraoral swelling is uncommon.

Clinical presentation

The periodontal abscess tends to have a sudden onset and there is usually swelling, redness and tenderness of the gingiva overlying the abscess. Pain may be continuous or related to biting and can usually be elicited clinically by percussion on the affected tooth. The periodontal abscess has no specific radiographic features although it is commonly associated with a deep periodontal pocket.

Collection of specimens

It is extremely difficult to collect a sample of pus from a periodontal abscess without contamination with subgingival plaque. As a result, samples are not routinely collected. However, if a microbiological investigation is required, e.g. in severe or recurrent infection both in a normal or compromised patient, the following general principles should be observed. The sample site should be isolated, supragingival and where possible subgingival plaque removed, and the supragingival area disinfected with chlorhexidine or povidone iodine. Light digital pressure is applied to the external wall of the abscess, with or without gentle subgingival instrumentation, until pus exudes from the mouth of the periodontal pocket. The sample should be collected on a sterile absorbant paper point held with sterile tweezers, transferred to anaerobic transport medium and sent to the microbiology laboratory immediately.

Microbiology

As might be expected, subgingival plaque is the source of the microorganisms normally isolated from periodontal abscesses. The microorganisms commonly present are: (1) anaerobic Gram-negative rods, especially black pigmented *Bacteroides* species and fusobacteria; (2) streptococci, especially α-haemolytic streptococci and anaerobic streptococci; and (3) other microorganisms such as spirochaetes, *Capnocytophaga* spp. and *Actinomyces* spp.

Sequelae

If a periodontal abscess is untreated, severe destruction of periodontal tissues may follow which may result in tooth loss.

Treatment

A thorough clinical assessment of the patient should be made. If the prognosis is poor, due to advanced periodontitis or recurrent infection and it is unlikely that treatment will achieve functional periodontal tissues, then the tooth should be extracted. If the abscess is relatively small and localized, extraction may be carried out immediately but, otherwise, extraction should be postponed until acute infection has been controlled. Drainage of the abscess should be encouraged and gentle subgingival scaling performed to remove calculus and any foreign objects. Irrigation of the pocket with warm normal saline should be carried out and the patient advised to use hot saline mouthwashes. If pyrexia or cellulitis is present, antibiotics should be prescribed, penicillin, erythromycin or metronidazole being the drugs of choice.

Suppurative osteomyelitis of the jaws

Osteomyelitis of the jaws is a relatively rare condition. It may present as an acute or chronic infection depending on a variety of factors. By definition it is an inflammation of the medullary cavity alone but, usually at a later stage of the disease, infection extends into the cortical bone and the periosteum. A wide range of microorganisms have been associated with osteomyelitis of the jaws, some of which will be described later in this chapter, e.g. *A. israelii*, and others which will appear in Chapter 8, e.g. *T. pallidum* and *M. tuberculosis*.

Aetiology

The source of infection in osteomyelitis of the jaws is derived from a contiguous focus or by haematogenous spread, the former being by far the more common. However, osteomyelitis is much rarer than dentoalveolar infections and this low incidence is probably due to the efficiency of the host defence mechanisms, especially the vascularity of bone. A number of conditions tend to reduce the vascularity of bone, e.g. radiation therapy, osteoporosis, Paget's disease, fibrous dysplasia and bone tumours, and thus compromise the host's defence mechanisms and so predispose to osteomyelitis. Osteomyelitis affects the mandible much more commonly than the maxilla and this is probably related to a number of anatomical differences, e.g. the relative paucity of medullary spaces and thin cortical plates found in the maxilla compared with the mandible.

Once within the medullary cavity, microorganisms proliferate and elicit an acute inflammatory reaction which results in an increase in intramedullary pressure, with pus formation, venous stasis and ischaemia. Pus spreads through the medulla and also via the haversian and nutrient canuli and, as a result, compromises the blood supply further by accumulating beneath the periosteum and elevating it from the cortex. Compression on nerves may produce anaesthesia. If the condition is untreated, the accumulation of pus reaches a stage where the periosteum is breached and sinuses result, with the appearance of soft tissue abscesses related to the oral mucosa or skin. If the disease is untreated, chronic osteomyelitis may develop with new bone formation occurring at the periphery of the area of destruction. This process separates fragments of necrotic bone (sequestra) from viable bone. In severe cases, areas of necrotic bone can become surrounded by a sheath of new bone. This is termed an involucrum.

Clinical presentation

The main clinical features of early acute suppurative osteomyelitis are pain, mild fever, paraesthesia or anaesthesia of the lower lip due to pressure on the mental nerve, and clear signs of the source of infection. As the condition progresses, the following additional signs and symptoms may appear: constant high temperature; loosening of teeth which are sensitive to percussion; exudation of pus from the gingival margin of affected teeth; or drainage of pus through mucosal or skin fistulae. The regional lymph nodes are usually enlarged and trismus and cellulitis may be present. In some cases of acute osteomyelitis, systemic upsets are minimal. In chronic secondary osteomyelitis, sinuses are usually present and the surrounding tissues are tender and indurated. The patient has little systemic upset.

In the early stages of acute osteomyelitis, radiographs will show no abnormality but by about two or three weeks after the start of the disease evidence of bone destruction ('moth eaten' appearance) can be seen.

Source and type of microorganisms

Since the majority of cases of osteomyelitis are related to existing dentoalveolar infection, the source of the causative microorganisms is the oral cavity and the species involved are similar to those shown in Table 6.1. However, there is some confusion in the literature concerning the role played by *Staphylococcus aureus* in osteomyelitis of the jaws. In older reports *Staphylococcus aureus* or *Staphylococcus epidermidis* (*albus*) were believed to account for about 80–90% of cases but, in more recent papers, anaerobic Gram-negative rods, especially *Bacteroides* species, are the predominant organisms, with *Staphylococcus* species being present in a minority of cases. The explanation for this change is not clear but it may be due to unsubstantiated comparisons with osteomyelitis in long bones, and failure to carry out adequate anaerobic cultures. The fact that *Staphylococcus aureus* is infrequently present in small numbers in the normal oral flora would tend to agree with this conclusion. Although almost any member of the oral flora may be isolated from cases of osteomyelitis of the jaws, it is usually polymicrobial, the common microorganisms isolated being pigmented *Bacteroides* species, fusobacteria and anaerobic streptococci. In some cases aerobic Gram-negative bacilli, e.g. *Klebsiella* species, *Proteus* species and *Pseudomonas aeruginosa*, can also be found. Since a wide range of microorganisms are likely to be involved, it is important to carry out microbiological investigations and perform antibiotic sensitivity tests to ensure that the most effective antibiotic is used.

Laboratory investigation

In the early stages of acute osteomyelitis it can be very difficult to obtain a satisfactory sample for microbiological examination and therefore empirical antibiotic therapy is employed. In the later stages of the disease, when pus exudes from the gingival margins or can be expressed from intraoral or extraoral sinuses, samples should be collected. In order to prevent the contamination of pus with saliva, strict aseptic collection techniques are required, the sample being collected on a sterile absorbent paper point which is placed directly into anaerobic transport media. If on examination the clinician suspects the presence of an area from which pus may be aspirated, the sample should be collected by that means and sent rapidly to the laboratory.

Treatment

The treatment of osteomyelitis is complex and requires to be organized on a personal basis for each case. The main stages in treatment are rapid diagnosis of the disease and the empirical prescription of antibiotics, since this may prevent further bone destruction and produce resolution of the condition without the need for surgical intervention. If possible, a pus sample should be collected and laboratory investigations carried out. The drugs of choice are penicillin G and, in allergic patients, clindamycin. Since in a proportion of cases *Staphylococcus aureus* is the causative organism and it is likely that the isolate will be β-lactamase positive, flucloxacillin may be required. Depending on the severity of the infection, further treatment options include tooth extraction, sequestrectomy and resection and reconstruction of the jaws.

Staphylococcal submandibular lymphadenitis in children

Aetiology

The disease is endogenous and caused by *Staphylococcus aureus*. By carrying out bacteriophage typing on staphylococcal isolates from the nose and abscess in the same patient, it has been shown that the anterior nares is the source of the causative organism. It is believed that *Staphylococcus aureus* spreads from the anterior nares to the submandibular lymph nodes via interconnecting lymphatics and, for reasons unknown, the organisms then proliferate and produce an abscess in that site.

Clinical presentation

The disease is a rare condition which is usually seen in children between 2 and 5 years of age, although younger and older children may be affected. The patient presents with a tender, erythematous, unilateral, submandibular swelling, occasionally with mild or severe trismus. Pain is not a marked feature, nor is severe constitutional upset. The total absence of any dental focus of infection is common. Many patients give a past history of recent antibiotic therapy (usually penicillin) prescribed by their medical practitioner.

Treatment

If the patient presents at an early stage in the infection, conservative treatment is indicated using flucloxacillin or erythromycin, since most strains of *Staphylococcus aureus* are β-lactamase producers. In more severe infections surgical drainage under general anaesthesia is indicated, followed by a short course of antibiotic therapy usually using flucloxacillin or erythromycin. Due to antibiotic resistance of the causative organism, cases of staphylococcal lymphadenitis should not be treated with penicillin or ampicillin since there is a danger of a chronically inflamed lesion developing with a thick fibrous capsule which can cause unnecessary delay in healing.

Cervicofacial actinomycosis

Actinomycosis in man may occur in a number of sites, such as the abdomen, lung or skin, but the cervicofacial region is the most common and accounts for well over half the recorded cases. The disease is a specific granulomatous process which is endogenous and characterized in the cervicofacial area by the presence of swelling which may be localized or diffuse and, if untreated, by the formation of discharging sinuses. Classically, this discharge of pus contains visible granules which may be gritty to touch and are known as 'sulphur granules'. This term is purely descriptive and sulphur is not the main constituent of the granules. The recognition of these granules in pus by a clinician may be the first indication that the infection which he is treating is actinomycosis.

Aetiology

In man, the infecting microorganism is *Actinomyces israelii* which is a common inhabitant of the oral cavity and may be isolated from plaque, carious dentine, calculus and debris from the tonsillar crypt. However, the relative paucity of cases of actinomycosis implies that factors other than the presence of the bacterium are necessary to produce the disease. It is thought that trauma to the jaws (most commonly due to tooth extraction or jaw fracture), or the presence of teeth with gangrenous pulps, are aetiological factors. The common routes of infection are either as a result of calculus or plaque becoming impacted in the depths of a tooth socket at the time of extraction, or by spread from a carious lesion via the pulp chamber and root canal of a non-vital tooth. However, these conditions occur much more commonly than actinomycosis and the other predisposing factors are unknown.

Clinical presentation

Actinomycosis is predominantly a disease of younger people, although all ages may be affected, and is more common in males than females. The infection can present in an acute, subacute or chronic form. Most cases of cervicofacial actinomycosis probably start as acute swellings which are indistinguishable on clinical grounds from acute dentoalveolar abscesses. The chronic form of the disease usually follows acute infection which has received no therapy or has been inadequately treated. The chronic form of the disease may also follow a subacute infection related to trauma.

Swelling is a common feature of presentation in actinomycosis and most commonly occurs in the submandibular region, although other sites such as the submental area and cheek may be involved. Cases have also been reported in the maxillary antrum, salivary glands, and tongue, as well as in localized intraoral situations such as periodontal abscess, pulp, apical granuloma and odontogenic cyst. Direct involvement of bone causing actinomycotic osteomyelitis also occurs. Pain is a variable feature and in some patients with acute infections, swelling is painless. A useful additional clinical feature of the painless group is that the swelling is consistently found at the point where facial vessels cross on the lower border of the mandible.

Other features which may be encountered, depending on the site of infection, are discharging sinuses, trismus, pyrexia, fibrosis around the swelling and the presence of infected teeth. Multiple discharging sinuses in the neck surrounded by areas of brawny induration are associated with chronic actinomycosis. Many patients with actinomycosis have histories of inadequate antibiotic treatment and this may greatly prolong the duration of the disease, due to repeated recurrence.

Source and type of microorganisms

Actinomyces israelii is the causative organism, although there are reports where other *Actinomyces* species, e.g. *Actinomyces naeslundii*, have been isolated from patients presenting with the typical signs and symptoms of the disease. In a minority of cases, *Actinobacillus actinomycetemcomitans* may be isolated in mixed culture with *A. israelii*.

Sampling and laboratory investigation

The first stage in the laboratory diagnosis of actinomycosis is to examine pus for the presence of 'sulphur granules'. If a fluctuant abscess is present, fluid pus should be collected by aspiration using a syringe, or in a sterile container if drainage by external incision is performed. A small drop of pus on the end of a dry cotton wool swab is virtually useless. In cases of chronic infection, a gauze dressing which has covered a discharging sinus for some time may be useful as a specimen. When the specimen arrives in the laboratory, it is usually necessary to dilute the pus in order that a search for 'sulphur granules' can be made. Gram films are made from any part with a lumpy or granular appearance. 'Sulphur granules' in acute cases of actinomycosis tend to be soft in consistency and can be easily mistaken for inspissated pus. The granules in chronic infection tend to be of a gritty consistency due to partial calcification. The classical microscopical appearance of a 'sulphur granule' stained by Gram's method is seen in Figure 6.4. Granules are washed,

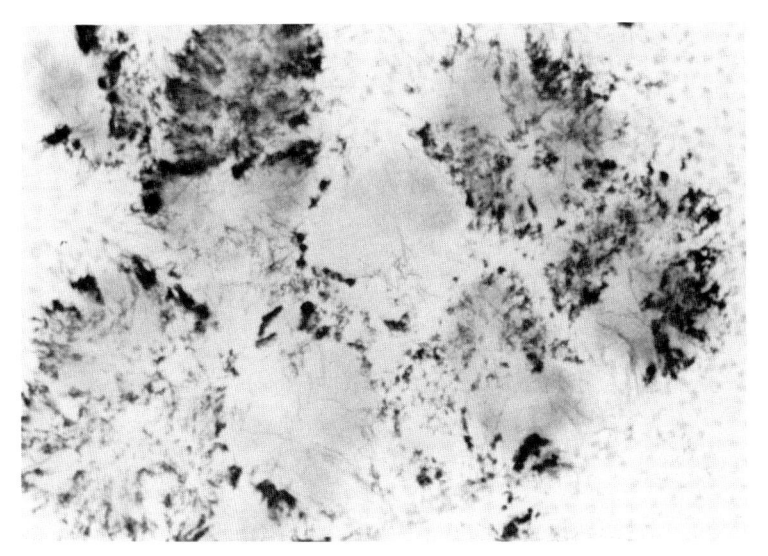

Figure 6.4 Gram-stained histological section of a 'sulphur granule' surrounded by pus cells. At the periphery of the granule, Gram-positive filaments can be seen (Magnification × 400, reduced to 60% on reproduction)

crushed in tissue grinders and cultured on blood agar plates under anaerobic conditions at 37 °C for 7 days. Colonies often produce a typical 'molar tooth' morphology. Pure cultures of suspected colonies are prepared and identified using biochemical techniques. A definitive diagnosis of actinomycosis requires the isolation of the causative organism and the appearance in pus of moderate to large clumps of Gram-positive branching filaments.

Treatment

A different prognosis and treatment has been indicated for acute and chronic forms of actinomycosis. The acute type shows, in general, a rapid resolution following the extraction of any associated dental focus, incision and drainage of facial abscess and a short (2 to 3 weeks) course of antibiotics, penicillin being the drug of choice. Subacute or chronic lesions require similar surgical intervention but a longer antibiotic course, 5–6 weeks on average, should be prescribed. If penicillin cannot be given, erythromycin and tetracycline are good alternatives, while some prefer clindamycin as it penetrates bony tissues well.

Further reading

Bronner, M. and Bronner, M. (1971) *Actinomycosis*, 2nd edn, Wright, Bristol

Lewis, M. A. O., MacFarlane, T. W. and McGowan, D. A. (1986) Quantitative bacteriology of acute dentoalveolar abscesses. *Journal of Medical Microbiology*, **21**, 101–104

Lewis, M. A. O., MacFarlane, T. W. and McGowan, D. A. (1988) Assessment of the pathogenicity of bacterial species isolated from acute dentoalveolar abscess using an animal model. *Journal of Medical Microbiology*, **27**, 109–116

Patterson, H. C., Kelley, J. H. and Strome, M. (1982) Ludwig's angina: an update. *Laryngoscope*, **92**, 370–378

Stenhouse, D., MacFarlane, T. W. and Still, D. M. (1978) Staphylococcal submandibular lymphadenitis in childhood. *British Journal of Oral Surgery*, **15**, 73–77

Chapter 7
Microbiology in endodontic therapy

The most common diseases which affect the dental pulp and periapical tissues are caused by bacteria. As a result, it is important for dentists to understand the pathogenesis of infections at these sites, since the application of knowledge is important in deciding on the form of treatment required for individual patients.

Source and route of infection

The sources and routes of infection which involve pulp and periapical tissues are: (1) carious dentine via dentinal tubules; (2) subgingival plaque bacteria present in deep periodontal pockets via lateral canals normally present in the apical two thirds of teeth; (3) from the oral environment secondary to tooth fracture or mechanical exposure; and (4) anachoresis (a mechanism whereby diseased tissue attracts bacteria present in bacteraemias). The importance of this last source is doubtful.

Pathogenesis of pulp and periapical infections

The clinical presentation and severity of these infections are related to the way in which the host defences interact with the infective agents present in the pulp cavity. An overall scheme of the stages involved in the spread of microorganisms from the pulp into the periapical tissues and possible sequelae are shown in Figure 7.1. Once bacteria gain access and proliferate in the pulp, it is unusual for the host defences to completely eliminate them. As a result, healing of infected pulp tissue is uncommon. The time necessary for complete necrosis of the pulp to take place is very variable; localized chronic infection may be present for many months or even years before the pulp dies. During this process the pulp may become sterile, especially if the carious cavity has been restored and the apical foramen occluded by calcified tissue. Usually viable microorganisms can be isolated from necrotic pulp tissue and they may either remain localized in the root canal or spread into the periapical tissues via the apical foramen or lateral root canals. It is not entirely clear if periapical granulomas develop in response to the presence of viable bacteria in the periapical tissues or if they form as a reaction to sterile, toxic products leaking from necrotic pulp tissue via the periapical foramen. However, recent studies suggest that the majority of granulomas contain relatively small

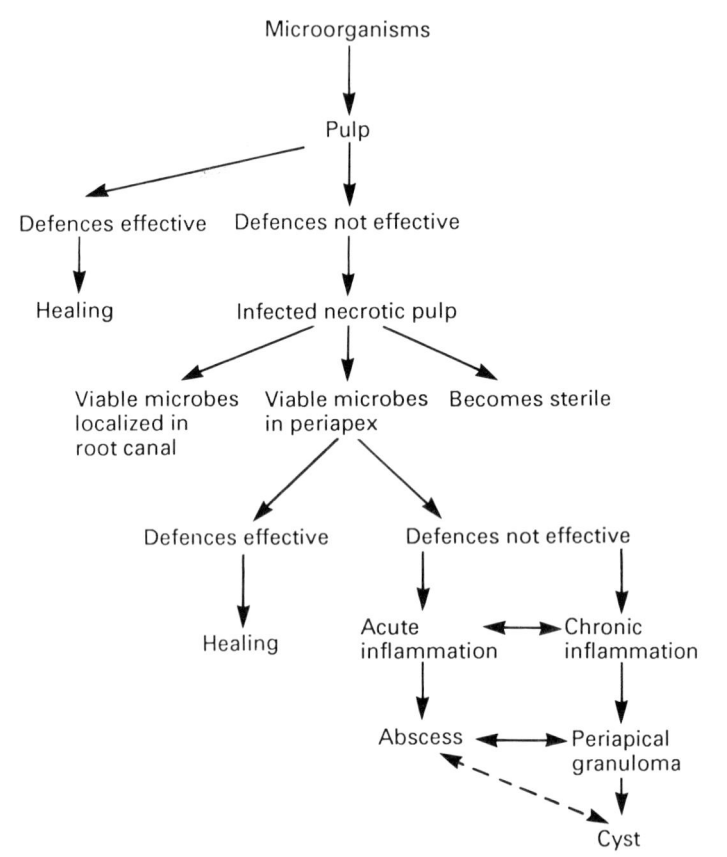

Figure 7.1 Stages in the development of infective lesions of the pulp and periapical tissues

numbers of bacteria which appear to be located within small areas of necrosis randomly scattered throughout the lesion. Little is known about the role, if any, of microorganisms in the development of periapical cysts. Dental cysts without any obvious pathway to the oral environment can become infected and it is possible that, in these cases, bacteria within the cyst itself proliferate due to alteration in the host–parasite relationship.

Host defence mechanisms

The initial defence mechanism of the pulp to infection is to prevent access of microorganisms to host tissues by depositing sclerotic or irregular secondary dentine in their pathway. Within the pulp, acute inflammation with phagocytosis may also occur, especially in the early stages of invasion, and later a chronic response may supervene with associated antibody-mediated immune reactions. In long-standing infections, cell-mediated responses may also participate. Since the pulp is enclosed in a mineralized shell and its blood supply usually passes through a single narrow foramen, it is likely that if extensive inflammation develops, widespread necrosis will rapidly occur due to a combination of stasis of the blood supply by compressive forces on blood vessels and release of lytic enzymes by polymorphonuclear leucocytes. Once the pulp is dead, the host defences have little influence on subsequent microbial interactions within the root canal.

Microbiology

The main groups of bacteria related to pulp and dentoalveolar infections are shown in Table 7.1. Generally the microflora associated with these infections are similar but different to that present in carious dentine. Since carious dentine is believed to be the main source of bacteria which infect necrotic pulp tissue, it is surprising that such differences occur. One likely explanation is that over a period of time complex interactions occur between the different species which enter the pulp via carious dentine. Perhaps with time the environment of the necrotic pulp may favour the very small numbers of strictly anaerobic bacteria derived from dentine and thus allow them subsequently to proliferate, and predominate.

Table 7.1 Predominant bacteria isolated from carious dentine, necrotic pulp tissue, periapical granuloma and dentoalveolar abscess

Microorganism	Carious dentine	Necrotic pulp tissue	Periapical granuloma	Dentoalveolar abscess
Facultative streptococci	2	2	3	2
Lactobacillus/ Actinomyces	4	2	2	1
Anaerobic streptococci	0	3	1	3
Fusobacterium spp.	0	2	1	1
Bacteroides spp.	0	3	2	3

0, not usually found; 1, occasionally present; 2, usually present in small numbers; 3, usually present in moderate numbers; 4, usually present in large numbers.

Strep. milleri is commonly isolated from endodontic infections, while black pigmented bacteroides, especially *B. endodontalis* and *B. gingivalis* are the predominant anaerobic Gram-negative bacili. However, representatives of almost all plaque microorganisms have been isolated from endodontic samples at one time or another. If a strict aseptic sampling method and good anaerobic culture techniques are used, strictly anaerobic bacteria can usually be isolated from endodontic samples from patients who experience infective problems during endodontic treatment. Infections related to endodontic therapy are usually polymicrobial (2–4 different species being involved), although monoinfections also occur.

It is interesting to note that when sudden flare-ups occur in symptomless, non-vital teeth with periapical lesions during 12–24 hours of starting endodontic treatment, anaerobes are rarely isolated and facultative Gram-positive cocci, mainly *Strep. salivarius, Strep. mutans* and enterococci, predominate.

The outcome of infection of the pulp by microorganisms is difficult to predict clinically and will depend on the net result of a range of interacting factors: (1) the source, route and duration of infection; (2) the species, number and toxic end-products of microorganisms; and (3) the status of the specific and non-specific defence mechanisms of the host tissues.

Role of microbiology in endodontics

It is generally accepted that the role of endodontic therapy is to render the root canal and periapical tissues 'sterile' and thus ensure the success of subsequent root canal filling. At first sight sterility testing is most attractive since it encourages a meticulous clinical technique and assists the clinician in deciding when the root should be filled. However, the validity of the term 'sterile' in the context of endodontics has come into some dispute. Sterility is an absolute term and indicates freedom from microorganisms. Thus if even one bacterial cell is present within a root canal then the canal cannot be termed sterile. Furthermore, there is good evidence that the presence of small numbers of bacteria in root canals is compatible with clinical success. This is not surprising since it is very unlikely that a few bacteria sealed within a well-prepared and filled root canal will be able to proliferate, release toxic products or spread into adjacent tissue. In addition there are many practical difficulties, mainly related to sampling and culture, which can produce false positive or negative results in sterility tests of root canals. Some of these factors are shown in Table 7.2. Thus for a number of reasons the case for routinely culturing samples from root canals is poor.

Table 7.2 Causes for false positive and negative results in sterility testing during endodontic treatment

False positive	*False negative*

Contamination of:
  root canal during treatment
  paper point during sampling
  cultures in laboratory
	Complex anatomy of root canal with associated difficulty in accurate sampling
	Poor laboratory methods
	Antimicrobial agents contaminating sample

However there are a number of clinical situations where the microbiological assessment of root canals can be justified. They are as follows: (1) patients suffering from dentoalveolar abscess where endodontic treatment is indicated; (2) symptomless non-vital teeth with apical radiolucency; and (3) root canals under treatment with obvious persistent exudate and clinical symptoms. On occasions root canals are left open to encourage pus from a dentoalveolar abscess to drain into the mouth. Wherever possible this should be avoided since a complex mixture of oral microorganisms selectively colonize the canal, and often prove difficult to eliminate (especially from the periapical area), during endodontic treatment. The case for performing microbiological tests on samples of necrotic pulp collected from symptomless, non-vital teeth which have a periapical radiolucency is that these canals are invariably infected. If the microorganisms present are characterized and their antibiotic sensitivity assessed at an early stage of treatment, then any infection problems which develop subsequently can be managed using rational rather than empirical therapy.

Microbiological sampling from root canals

Samples from root canals should be collected on sterile paper points, using a strict aseptic technique. Failure to use such a technique may well lead to contamination of the sample and make interpretation of the microbiological findings difficult. If the canal is dry either the paper point or the canal should be moistened with sterile fluid before sampling. After sampling, the paper point is placed in an anaerobic transport medium which is then sent directly to the laboratory. The sample is dispersed by vortex mixing and a standard inoculum is cultured on blood agar which enables a count of microorganisms/ml medium to be calculated. The microorganisms present in the sample are identified and tests performed to assess their sensitivity to antimicrobial agents. In addition to providing a simple positive or negative result, it is possible by this means to give the clinician both qualitative and quantitative information about the microorganisms present in the root canal, as well as guidance about the overall antibiotic sensitivity results.

Treatment

The most important part of endodontic therapy is the use of aseptic technique to remove mechanically vital, non-vital or infected tissues, and to prepare the canal so that complete root canal filling can be subsequently performed. In the past, antimicrobial agents were commonly used as an adjunct to endodontic therapy but in recent times their use has declined and some clinicians believe that they should be used only in the management of cases complicated by severe or persistent infection. Antimicrobial agents can be used in three main ways: (1) as irrigants to wash out canals during mechanical cleansing procedures; (2) as topical agents sealed in the root canal for a few days to kill microorganisms inaccessible to mechanical therapy; and (3) as systemic agents to destroy microorganisms within the periapical tissues. Many of the topical agents used are volatile and it has been suggested that in vapour form they can penetrate into the periapical tissues and destroy microorganisms.

This appears unlikely and antibiotics should be used if the clinician is attempting to control infection which he believes to be in the periapical tissues. A wide range of antimicrobial agents have been used in endodontic therapy over the years; a few of the common agents are shown in Table 7.3. From a microbiological point of view there seems little sense in using antiseptics routinely, either as irrigants or as topical medicaments, unless there is some evidence that infection is, or is likely to become, a problem. Many antiseptics are inactivated by blood and serum (which may enter the canal from the periapical tissues) and may therefore fail to exert any lasting effect on the microflora either within the canal or perapical tissues. Furthermore, many medicaments are toxic to host tissues and if they pass from the root canal into the periapex they are likely to cause inflammation or necrosis. Depending on the severity of damage and whether or not nerves are affected, the patient may experience pain or loss of sensation. Mild irritation to frank necrosis of the soft tissues and bone in the periapical tissues can cccur, with subsequent delay in healing. There is a strong case for discontinuing the use of most medicaments, especially those containing paraformaldehyde. On balance it is likely that the tissue damage which can result from the use of disinfectants in endodontics far outweighs their mild antimicrobial activity when used *in vivo*. Therefore, whenever possible, substances with low or zero toxicity should be used as a root canal irrigant, e.g. sterile normal saline.

Table 7.3 **Antimicrobial agents used in endodontics**

Irrigants	*Topical agents*	*Systemic drugs*
Sodium hypochlorite	Beechwood creosote	Penicillin
Hydrogen peroxide	Formocresol	Amoxycillin
Iodine compounds	Paraformaldehyde	Metronidazole
Normal saline	Noxythiolin	Erythromycin

It is difficult to give definitive guidance on the use of antimicrobial agents in endodontics since much depends on the experience and skill of the operator. Antibiotics should be prescribed in the management of periapical infections, certainly in the acute phase, and arguably in chronic but persistent infections which cannot be controlled by any other means. It is important to note that there is no case for using antimicrobial agents as an adjunct to inadequate and careless clinical technique.

Further reading

Bystrom, A. and Sundqvist, G. (1981) Bacteriologic evaluation of the efficacy of mechanical root canal instrumentation in endodontic therapy. *Scandinavian Journal of Dental Research*, **89**, 321–328

Fabricius, L., Dahlen, G., Holm, S. E. and Moller, A. J. R. (1982) Influence of combinations of oral bacteria on periapical tissues of monkeys. *Scandinavian Journal of Dental Research*, **90**, 200–206

Morse, D. R. (1987) Microbiology and pharmacology. In *Pathways of the Pulp* (eds S. Cohen and R. C. Burns), 4th edn, Mosby, St. Louis, pp. 364–396

Sundqvist, G. K., Eckerbom, M. I., Larson, A. P. and Sjogren, U. T. (1979) Capacity of anaerobic bacteria from necrotic pulps to induce purulent infections. *Infection and Immunity*, **25**, 685–693

Van Steenbergen, T. J. M., Van Winkelhoff, A. J., Mayrand, D. et al. (1984) *Bacteroides endodontalis* sp. nov. an asaccharolytic black pigmented bacteroides species from infected root canals. *International Journal of Systematic Bacteriology*, **34**, 118–120

Part 3
Infections involving the oral and perioral tissues

Chapter 8
Bacterial infections

Specific bacterial infections of the mouth are uncommon, with tonsillitis in children caused by *Strep. pyogenes* being the most common. Other bacterial causes of pharyngitis which are rarely diagnosed today include *C. diphtheriae* and the fuso–spirochaetal complex associated with Vincent's angina. Specific infections of the oral mucosa which are rare, and are usually manifestations of systemic infections, include diseases such as tuberculosis, syphilis and gonorrhoea. Although many of these infections are uncommon in North America and Europe, the opposite is true in many other parts of the world.

Tonsillitis and pharyngitis

Sore throat is a common symptom of a number of specific infections and is accompanied by a variable degree of constitutional upset. Although a specific diagnosis may be assisted by the clinical presentation, it is often necessary to perform laboratory tests. At least two-thirds of these infections are caused by viruses, the remaining one-third by bacteria, almost all due in the developed countries to *Strep. pyogenes*. It is noteworthy that *Strep. pyogenes* may cause rheumatic fever and carditis in susceptible individuals. Consequently, they may be at a higher risk of contracting infective endocarditis due to bacteraemias associated with surgical procedures such as tooth extractions. A summary of the clinical presentation, causative agents and microbial diagnosis of tonsillitis/pharyngitis is shown in Table 8.1.

Gonorrhoea

General

Gonorrhoea is a venereal infection with a worldwide distribution which has reached epidemic proportions in some countries. The

Table 8.1 Bacterial causes of tonsillitis – pharyngitis

Causative agent	Clinical presentation	Sequelae and late complications if not treated	Laboratory diagnosis	Treatment
Streptococcus pyogenes (Lancefield Gp. A) Gram-positive cocci, 1 μm, in chains	Acutely inflamed tonsils with or without pus in crypts	Peritonsillar abscess, sinusitis, otitis media, scarlet fever, rheumatic fever and acute glomerulonephritis	Swab or saliva sample. Culture on blood agar, β-haemolytic colonies, Lancefield grouping and identification using API strep.	Penicillin, or erythromycin if hypersensitive
Corynebacterium diphtheriae Gram-positive rods 3 × 0.3 μm, 'Chinese character' arrangement with metachromatic granules	Inflamed fauces covered with grey-white serocellular membrane	Cranial and peripheral nerve palsies. Death due to (1) respiratory obstruction (mechanical blockage by membrane) (2) heart failure (cardiotoxin)	Swab of membrane and culture on blood tellurite and Loefflers medium. Identification by biochemical tests and toxin production (Elek plate)	Antitoxin, and tracheostomy if required. Penicillin or erythromycin
Fuso-spirochaetal complex, *Treponema* species and *Fusobacterium nucleatum* (see Chapter 5)	Necrotic ulcers on tonsils	Possibly chronic infection	Smear from ulcer stained by Gram's method	Penicillin or metronidazole

causative agent *Neisseria gonorrhoeae*, a Gram-negative diplococcus, is transmitted by sexual activity, and lesions occur in the genitals, the anal canal, the oral cavity (particularly the pharynx), and combinations of these sites. The major risk factor for gonococcal pharyngitis is practice of orogenital sex. Gonococcal pharyngitis can be found in 10–20% of heterosexual women, 10–25% of homosexual men and 3–7% of heterosexual men who have urogenital gonorrhoea.

The disease is caused by the adherence of *N. gonorrhoeae* to either the genital or oral mucosa, and subsequent invasion into the deeper tissues between the surface epithelial cells. Spread may then occur by direct continuity, by lymphatics or haematogenously. Locally there is an intense inflammatory reaction with the production of pus which on microscopy reveals Gram-negative diplococci (0.6–1.0 μm), both extracellularly and within a proportion of polymorphonuclear leucocytes.

Oral manifestations

The oral mucosa is relatively resistant to infection by *N. gonorrhoeae* but gonorrhoeal stomatitis is diagnosed with increasing regularity in sexually active adults, especially homosexuals. The oral lesions of gonorrhoea are more commonly associated with primary infection of the mouth with *N. gonorrhoeae* than with spread of the pathogens to the mouth from a distant site.

The clinical presentation of primary oral gonorrhoea is varied, depending on the severity and distribution of the infection. The patient usually complains of a burning or itchy sensation or a dry hot feeling in the mouth, which in 24–48 hours changes to acute pain. The patient may also complain of a foul oral taste, the salivary flow may be increased or decreased, the breath may be foetid and the submandibular lymph nodes are usually enlarged and painful. The temperature may be raised if the infection is severe. Lesions have been reported on all parts of the oral mucosa and are usually described as consisting of a variable mixture of the following clinical signs: inflammation, oedema, vesiculation, ulceration, and pseudomembranes. These last are white, yellow or grey in colour, and easily removed by scraping which leaves a bleeding surface. In some patients a diffuse painful stomatitis occurs in which mucous membranes become fiery-red and oedematous. Speech, swallowing and simple mouth movements become extremely painful. The oral sites of primary infection which have been reported are the gingivae, the tongue, the buccal mucosa, the hard and soft palates and the oropharynx. *N. gonorrhoeae* have also been isolated from lesions of central papillary atrophy of the tongue. The tonsils and the oropharynx appear to be the commonest site of oral infection. However, as oral mucosa anterior to the pharynx is considered to be highly resistant to gonococcal infection, it is now thought that at least some of the previous reports of gonorrhoea in these locations are equivocal.

There is evidence that primary infection of the mouth can lead to disseminated lesions. Secondary gonorrhoeal infection of the parotid gland is rare but may occur in patients who have oral gonorrhoea or carry the organisms in their oropharynx. The significance of persistent oropharyngeal carriage of gonococci in asymptomatic individuals is not clear. Gonococcal arthritis of the temporomandibular joint is a not uncommon sequel if the primary infection is in the pharynx.

Gonorrhoeal infection of the urogenital tracts may result in haematogenous spread of gonococci to the skin and oral mucosa with resulting secondary infection. The oral manifestations of this type of infection can be split into two main groups, both of which appear to be rare: septic embolic phenomena and hypersensitivity reactions.

The recognition and diagnosis of oropharyngeal gonorrhoea is important for three main reasons: firstly, cases may act as a reservoir for the transmission of infection; secondly, they are a potential source of gonococcaemia giving rise to arthritis, meningitis and other septic embolic lesions in the body; and thirdly, they cause severe local tissue damage and discomfort.

Laboratory diagnosis

The only way in which a diagnosis of gonorrhoeal stomatitis can be made is by laboratory investigation.

Direct examination
The examination of Gram-stained films from oral lesions are of little direct diagnostic value due to the presence of oral commensal *Neisseria* species which cannot be differentiated from *N. gonorrhoeae* microscopically. The presence of numerous Gram-negative intracellular diplococci in a smear from a suspect lesion should be investigated further.

Culture
Swabs taken from the oral lesions should be placed in bacter-

iological transport medium and sent to the laboratory as quickly as possible. The requirement for culture is a semiselective medium, e.g. Thayer–Martin, containing antimicrobial agents, e.g. lincomycin, colistin and nystatin, to inhibit the commensal flora. Incubation in 5–10% CO_2 at 37 °C for 2 days results in colonies which can be identified using oxidase test, carbohydrate utilization (acid from glucose only) or fluorescent tests.

Serology
There is no suitable serological test available for use in the routine diagnosis of gonorrhoeal infection at present, nor has a successful vaccine been produced.

Treatment

Standard treatment of 4.8 million units of procaine penicillin G, either by intramuscular injection alone or combined with 1.0 g probenecid orally, is effective for pharyngeal gonorrhoea. An alternative regimen is 0.5 g tetracycline orally four times a day for seven days. Amoxycillin or spectinomycin do not always eradicate pharyngeal infection.

Syphilis

Although syphilis is not encountered frequently today, it is an important disease with a variety of oral manifestations. At present in the West, it is most commonly seen among homosexuals. The causative organism *Treponema pallidum* is a motile spirochaete which can not be cultured *in vitro*. Spirochaetes have a helical structure which consists of a central protoplasmic cylinder bounded by a cytoplasmic membrane with an outer cell wall similar to that of Gram-negative bacteria. Beneath the outer membrane run axial filaments which can be regarded as internal flagella.

The course of syphilis when acquired after birth can be split into three main stages.

Primary acquired syphilis

After the initial exposure to infection with *Trep. pallidum*, the spirochaetes pass through the mucous membrane or skin and are then carried in the blood throughout the body. After an incubation period of about three to four weeks there develops, at the site of entry, an ulcerated lesion called a primary chancre. The ulcer is round to oval with a regular bevelled margin and a base which is indurated, smooth and brownish red. The typical lesion is painless and oedema of the surrounding tissues is usually present. The regional lymph nodes become enlarged about two weeks after the appearance of the chancre and, on examination, are firm, painless and discrete with a rubbery consistency.

Oral manifestations

The normal site for the chancre is on the genitalia but extragenital primary lesions occur in some 10% of patients with syphilis. A chancre of the lip is the most common extragenital lesion and may present at the angles of the mouth. Other sites affected are the tongue and, to a lesser extent, the gingivae and tonsillar area (Table 8.2). Intraoral chancres are usually slightly painful due to secondary bacterial infection. The lesions are infectious and transmission may occur by kissing, unusual sexual practices and even by intermediate contact with cups, glasses and eating utensils. The chancre heals spontaneously about one to five weeks after appearing. The regional lymph nodes (submaxillary, submental and cervical) are enlarged. The differential diagnosis of primary syphilis includes ruptured vesicles of herpes simplex, traumatic ulcers and carcinoma.

Table 8.2 Oral manifestations and infectivity of syphilis

Stage	Oral manifestations	Infectivity
Primary	Chancre of lip, tongue gingiva	+++
Secondary	Mucous patches on tonsils, tongue, soft palate, cheek. 'Snail track' ulcers. Rubbery, enlarged cervical lymph nodes	++
Tertiary	'Gumma' of palate. Rarely osteomyelitis. Syphilitic leukoplakia leading to carcinoma	±
Congenital	Hutchinson's incisors. 'Mulberry' molars. Facial deformities with open bite or dish face	−

+++, Very high; ++, high; ±, very low; −, nil.

Secondary acquired syphilis

Six weeks after healing, the secondary stage of syphilis usually develops, and in about one-third of patients the primary chancre may be present at the same time. The signs of secondary syphilis are variable: generalized skin lesions in about 75% of cases, mucosal ulcers in about 33%, generalized lymphadenopathy in about 50%, and systemic symptoms which are 'influenza like', including fever, headache, malaise and general aches and pains. The skin lesions are found predominantly on the face, hands, feet and genitalia, and appear as dull red macular or papular spots.

Oral manifestations

The oral manifestations are classically slightly raised, greyish-white, glistening patches on the mucous membrane of the tonsil, soft palate, tongue or cheek but rarely the gingival tissues (Table 8.2). Lesions which form on the larynx and pharynx may lead to hoarseness. The surface of these lesions is covered with a greyish membrane which is easily removed and contains many spirochaetes. The mucous patches may coalesce to produce a serpiginous lesion called a 'snail track ulcer', and the cervical lymph nodes are enlarged and rubbery in consistency. Maculopapular eruptions have been described in the mouth and are confined mainly to the palate but occasionally the entire oral mucosa may be involved. In recent years the secondary syphilitic lesions found in the mouth are often atypical, due, in many cases, to inadequate treatment as a result of antibiotic therapy for an unrelated infection.

The differential diagnosis of secondary syphilis includes aphthous ulcers, erythema multiforme, lichen planus and tonsillitis. The secondary stage lesions heal two to six weeks after the time they first appear.

Latent and tertiary syphilis

The disease then becomes latent and may last a lifetime, or the infection may become active as the tertiary stage of syphilis. Reactivation can occur at any time between three to ten years after the date of first infection but the lesions of this stage of syphilis are rarely seen today. The characteristic lesion is the gumma, which is usually localized, single or multiple, varying in size from a pinhead to a lesion several centimetres in diameter. Gummas develop in the skin, the mucous membranes, and bones. The other organs involved in tertiary syphilis are the cardiovascular system and the nervous system.

Oral manifestations
The most common site for gumma formation in the mouth is the hard palate, although the soft palate, the lips, tongue and face

are also relatively commonly involved (Table 8.2). The lesion starts as a small, pale, raised area which ulcerates and rapidly progresses to a large zone of necrosis with denudation of bone, and in the case of the palate may eventually perforate into the nasal cavity. The palatal lesions are usually midline and in rare cases the soft palate may be involved. Gummas are painless and have low infectivity.

Occasional cases of syphilitic osteomyelitis involving the mandible and, less commonly, the maxilla have been described. The condition is characterized by pain, swelling, suppuration and sequestration, and both clinically and radiographically it may resemble pyogenic osteomyelitis.

Atrophic or interstitial glossitis is another oral manifestation of tertiary syphilis. Clinically, there is atrophy of the filiform and fungiform papillae which results in a smooth, sometimes wrinkled, lingual surface. Loss of the protective papillae subjects the dorsum of the tongue to many noxious stimuli and leucoplakia frequently develops. There is some evidence that syphilitic glossitis is related to the development of carcinoma of the tongue but the relationship is not clear since most patients studied had other predisposing factors (Table 8.2).

Late and quaternary syphilis

Ten to twenty years after primary syphilis the quaternary stage may develop, characterized by two main clinical forms: cardiovascular and neurosyphilis. Those who develop general paresis of the insane consequent to neurosyphilis may have perioral tremors and fine irregular tremors of the tongue and fingers.

Congenital syphilis

Oral manifestations
The dental lesions of congenital syphilis are a result of infection of the developing tooth germ by *Trep. pallidum*. Since the deciduous teeth are usually well developed by the time the spirochaetes invade the developing dental tissues, these teeth are minimally affected. The permanent teeth are at an early stage of development and infection may result in the complete failure of development of a tooth or the production of malformed teeth. The most common dental manifestation of congenital syphilis is the so-called 'mulberry molar' teeth, which are highly suggestive of prenatal syphilis. The first permanent molar teeth are usually involved and have roughened dirty, yellow,

hypoplastic, occlusal surfaces with poorly developed cusps and are smaller in size than normal. Hutchinson's incisors are another manifestation of congenital syphilis. The upper central incisors are usually involved and have crescentic notches in the middle of their incisal edge. The crown tends to be wider gingivally than at the incisal edge, giving a 'screwdriver' appearance. The lower incisors show similar defects but they are affected much less often than their maxillary counterparts. Infection of the developing bones of the face may lead to permanent facial deformities which produce an open bite and a 'dished' appearance to the face (Table 8.2).

Laboratory diagnosis

Direct examination

Exudate from primary or secondary lesions may be collected in fine capillary tubes and examined using dark-ground microscopy for typically motile *Trep. pallidum* (i.e. slender filamentous helices, 6–14 μm × 0.2 μm with 6–12 evenly spaced coils). This technique has limited value in lesions affecting the mouth since the oral commensal *Trep. microdentium* closely resembles *Trep. pallidum*. They do not stain by Gram's method but, using a silver staining technique, e.g. Levaditis', they appear as black spiral cells. Specific immunofluorescence tests, if available, are helpful in diagnosis.

Serology

The laboratory diagnosis of syphilis is usually made by serology since *Trep. pallidum* cannot be routinely cultured *in vitro*. Ten millilitres of venous blood are sufficient to carry out all the serological tests for syphilis. It is usual to perform at least two

Table 8.3 Serological tests for the diagnosis of syphillis

Stage of disease	VDRL	TPHA	FTA Abs.
Primary	+ or −	+ or −	+
Secondary or tertiary	+	+	+
Late (quaternary)	+	+	+
Treated syphilis	−	+	+
Congenital	+	+	+

VDRL, Venereal Diseases Reference Laboratory; TPHA, *Treponema pallidum* haemagglutination; FTA Abs., fluorescent treponema antibody absorbed.

different tests on each specimen. Two types of antigens are available: (1) cardiolipin, not derived from spirochaetes, which is used in Venereal Diseases Reference Laboratory (VDRL) tests; and (2) specific antigens derived from *Trep. pallidum* which are used in *Treponema pallidum* haemagglutination (TPHA) and fluorescent treponema antibody absorbed (FTA Abs.) tests. A simplified account of the interpretation of these tests is shown in Table 8.3. It should be remembered that diagnosis and treatment of syphilis should be left to an expert in venereal infections.

Treatment

The most effective drug is procaine penicillin. Tetracycline or erythromycin can be used in patients who are hypersensitive to penicillin. Follow-up with regular clinical and serological examinations are necessary for at least two years.

Tuberculosis

Mycobacterium tuberculosis and *Mycobacterium bovis* cause, respectively, human-type and bovine-type tuberculosis in man. Other opportunist mycobacteria may also cause human infections.

Using Ziehl–Neelsen stain, mycobacteria appear as red, acid–alcohol-fast slender bacilli, 1×4 μm, occurring singly or in small clumps. They stain poorly with Gram stain due to the high lipid content of the cell wall.

Tuberculosis occurs worldwide and although the incidence is relatively low in Europe and North America it is estimated that about five million cases of tuberculosis occur in the world annually. Primary infection in humans usually occurs in the lungs and consists of a localized lesion at the site of impaction of the bacilli and, at a later stage, regional adenitis. Initially, little or no tissue reaction takes place but, after a period of weeks or months, delayed hypersensitivity to the tubercle bacillus develops and the host tissues react with an intense inflammatory response with caseation. The lesion may heal by fibrosis and calcification, or an equilibrium between the host and the tubercle bacilli may develop, which, if disrupted, may produce symptomatic disease many years later. The lesions may progress, spreading locally in the lung and eroding into bronchi, producing open tuberculosis, or into the bloodstream, to produce a disseminated form of the disease.

Oral manifestations

Oral lesions to tuberculosis are usually secondary to primary tuberculous infection elsewhere in the body, commonly in the lung. Primary infections of the oral mucosa by *Mycobacterium tuberculosis* have been described but they are rare. The sources of infection are infected sputum or blood-borne bacilli in the case of secondary infection, while in primary infection the source is unknown. Lesions are found more commonly in the posterior parts of the mouth and it has been suggested that this is related to the relative distribution of lymphoid tissue.

Oral ulceration
There is wide variation in the clinical presentation of tuberculous lesions of the oral mucosa, with ulceration and pain being commonly reported. However, diffuse inflammatory lesions, granulomas and fissures have also been described, and pain may be mild or absent. The tongue is probably most commonly affected but lesions have been noted on the buccal mucosa, gingivae, floor of the mouth, lips, and the hard and soft palate. Primary tuberculosis of the oral mucosa is more common in children and adolescents than adults and usually presents as a single painless indolent ulcer, commonly on the gingiva, with enlarged cervical lymph nodes, or as a white patch. A past history of mechanical trauma is not uncommon in many patients with oral tuberculosis, and extraction sockets, the sharp edges of teeth, dentures and various restorations have been implicated as sites of entry for *M. tuberculosis*. In a number of cases, biopsy of oral lesions has subsequently led to a diagnosis of respiratory tract tuberculosis.

Tuberculous lymphadenitis
The cervical group of glands is most commonly affected by tuberculous lymphadenitis and in patients with pulmonary tuberculosis the route of infection is probably by lymphatic or haematogenous spread, or via an abrasion of the mouth. In patients with no evidence of systemic infection, the route of infection is probably via the tonsils or oral mucosa, with or without a clinical lesion at the site of entry. The patient usually complains of a lump in the neck which is typically but not always painful. The size may vary from 5 or 6 mm to several

centimetres in diameter, and in the early stages the swelling is firm but mobile. Later the mass becomes fixed with the formation of an abscess and sinus. The lesions may be unilateral, bilateral, single or multiple. The differential diagnosis should include oral and dental infection, actinomycosis, sarcoidosis, reticulosis, bronchial carcinoma and leukaemia. Generally, atypical mycobacteria are commonly involved in cervical lymphadenitis in children, while *M. tuberculosis* is more likely in adults.

Periapical granuloma and bone infections
In patients with active tuberculosis, tubercle bacilli have been demonstrated in 8% of periapical granulomas. Tooth extraction may lead to delayed healing of the socket which fills with so-called tuberculous granulations. Bone infections are not uncommon in tuberculosis, and secondary tuberculous osteomyelitis may involve the maxilla or mandible. The tubercle bacilli may gain access to the bone by haematogenous spread, by direct spread from an oral lesion or by infected saliva entering an extraction socket or fracture. Cases of tuberculous osteomyelitis of the jaws are chronic in nature, usually with severe pain and the production of bony sequestra.

Tuberculosis of the salivary gland is discussed in Chapter 11.

Laboratory diagnosis

Direct examination
The numbers of tubercle bacilli present in oral mucosal lesions are usually very small and hence direct examination of scrapings stained with Ziehl–Neelsen stain are usually negative. Pus or necrotic debris in cases of tuberculous osteomyelitis, parotitis or lymphadenitis are more likely to prove positive on microscopical examination. Because of the difficulty in diagnosing oral tuberculous infection, biopsy is usually undertaken; ideally, part of the biopsy should be sent for culture and part fixed for histological evaluation.

Culture
Samples are inoculated on to Lowenstein–Jensen medium and incubated at 37 °C for about two to three months. Human and bovine strains of *M. tuberculosis* take about six to eight weeks to produce recognizable growth, while colonies of atypical mycobacteria can be recognized after two to seven days. In some cases, inoculation of guinea pigs may be necessary to

isolate *M. tuberculosis*. Atypical mycobacteria produce visible growth on Lowenstein–Jensen medium within a few days at a wide range of temperatures. Pigmented colonies are common.

Serology
The Mantoux test may be of value in the diagnosis of the oral manifestations of tuberculosis. A positive test indicates present or previous infection with *M. tuberculosis* or *M. bovis*, although in a patient with overwhelming infection, or one receiving immunosuppressive drugs, the test is negative. Differential tuberculin testing may be of help in diagnosing cervical lymphadenitis in children.

Radiology

Radiographic examination of the lungs and, if necessary, other areas of the body is required in all patients where there is an evidence of oral tuberculosis.

Treatment

Generally the treatment of the oral manifestations comprises the use of antituberculous drugs with or without surgical intervention. The standard chemotherapy for tuberculosis is a combination of isoniazid and ethambutol or rifampicin. Atypical mycobacteria may be resistant to normal antituberculous chemotherapy and in children with cervical lymphadenitis caused by this group of microorganisms surgical excision may be necessary.

Leprosy

Leprosy is caused by an acid–alcohol-fast bacillus, *Mycobacterium leprae*, which in general prefers to grow near the cooler body surfaces of man.

It has been estimated that about 12 million people in the world suffer from leprosy. Although most of the sufferers live in tropical areas, cases of leprosy occur in other countries, especially those bordering the Mediterranean, Adriatic and Black Seas. The disease is rare in Northern Europe and America and is usually found within the immigrant population from endemic countries. Droplet infection is the probable mode of transmission of leprosy, the source of the infection being the nasal discharge of patients with lepromatous leprosy. However, the disease is weakly infectious and prolonged contact is thought to be necessary to acquire infection.

Leprosy is a chronic infectious disease of peripheral nerves, skin and sometimes other tissues. *Mycobacterium leprae* usually causes a mild or subclinical infection on initial entry to the tissues. The subsequent clinical pattern of disease depends on the type and extent of the host immune reaction. Two main forms of leprosy are recognized, lepromatous and tuberculoid (Table 8.4). In addition a third category, borderline leprosy, has been described, in which patients progress from a tuberculoid-like form of the disease to a more lepromatous presentation.

Table 8.4 A comparison of the different types of leprosy

Type of disease	Cell-mediated immunity	Antibody response	Widespread lesions	Numbers of M. leprae in lesions
Tuberculoid	++	−	−	±
Lepromatous	− or ±	++	+	++

Oral manifestations

Tuberculoid leprosy

There is no record of any case of tuberculoid leprosy affecting the oral mucosa but the neurological features associated with this form of leprosy may affect the mouth and face. The changes which may be seen vary from loss of eyebrows and eyelashes to nodular involvement of all facial cutaneous and subcutaneous structures (Figure 8.1). If the trigeminal nerve is involved, hyperaesthesia or paraesthesia of the face, lips, tongue, palate, cheeks or gingiva may be present. Secondary ocular changes may also occur when the trigeminal nerve is affected with subsequent corneal and conjunctival sensory loss. The dermal lesions of tuberculoid leprosy on the face are normally single and, elsewhere, if multiple, have an asymmetrical distribution. The lesions consist of dry, hairless plaques, with a well-defined and raised border, which on white skin are red and on dark skin are hypopigmented. The patient usually complains that the lesion is numb and on testing is anaesthetic to touch and pinprick.

In borderline leprosy, oral lesions appear as focal or diffuse enanthemas or papules, rarely forming ulcers.

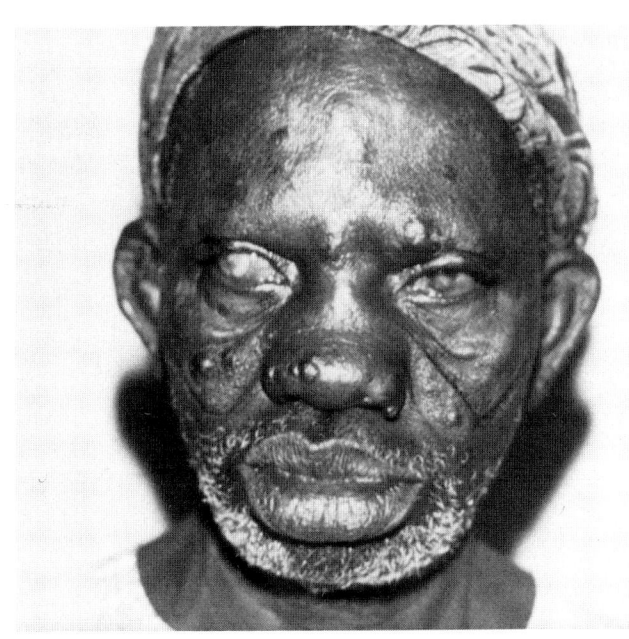

Figure 8.1 Leprosy in an elderly man manifesting saddle nose, sagging skin and blindness

Lepromatous leprosy
In lepromatous leprosy, *M. leprae* is present in many tissues of the body and multiple, erythematous, bilateral and symmetrical lesions are found on the skin of the face, arms and legs. The lesions are anaesthetic. Bone lesions also occur in the hands, feet and skull, resulting in the development of deformities in these areas. The nasomaxillary complex is the primary area of destruction in the facial region. Facial skeletal changes such as saddle nose, atrophy of the anterior nasal spine and premaxillary bone recession are frequently seen, with or without concomitant tooth loss. Dental deformities appear to be limited to a pink discoloration of the upper incisors, due to invasion of the pulp by infected granulomatous tissue which can produce pulpitis and pulp death.

The reported incidence of oral lesions in lepromatous leprosy vary from 12% to 58%. Intraoral nodules have been described as yellowish-red, soft-to-hard, sessile, single or confluent lesions which tend to ulcerate. Healing is by secondary intention with fibrous scars. The sites most commonly involved are the premaxillary gingivae, the hard and soft palates, the uvula and tongue. Tongue lesions, particularly on the anterior two thirds, occur in about 2–25% of patients: the lesions consist of single or multiple nodules giving a cobble-stone appearance or, in some instances, may resemble a geographic tongue. Since the saliva of patients with oral lesions of leprosy commonly contain *M. leprae*, it is likely that this could be an important source of infection.

Laboratory diagnosis

The diagnosis of leprosy is not usually difficult as long as the patient is carefully examined for the signs and symptoms of the disease, i.e. anaesthetic skin lesions and thickened peripheral nerves.

Microscopic examination

In lepromatous leprosy, an incision is made into the dermis and the lesion scraped using the blunt side of the blade. A smear is made and stained by the Ziehl–Neelsen method to demonstrate acid- and alcohol-fast bacilli. In tuberculoid leprosy it is difficult to demonstrate bacilli and the disease is diagnosed by histological examination of a biopsy of skin or thickened nerve.

Culture
M. leprae has never been cultured *in vitro* but experimental infection of armadillos and thymectomized irradiated mice has been successful.

Skin test
The lepromin test is of little diagnostic value.

Treatment

Treament should be undertaken by those with experience of leprosy and advice from the Panel of Leprosy Opinion is essential. Therapy consists of the administration of long-acting diaphenylsulphone (dapsone), with rifampicin plus clofazimine or ethionamide or prothionamide. Multi-drug regimens with bactericidal drugs have been recommended by the World Health Organization following worldwide resistance to dapsone.

Tetanus

Tetanus is caused by a strictly anaerobic spore-bearing Gram-positive ('drumstick') bacillus, *Clostridium tetani*. The spores of the organism germinate under anaerobic conditions and produce an exotoxin (tetanospasmin), as well as a haemolysin (tetanolysin) which reaches the brainstem and the anterior horn of the spinal cord, either between fibres of the regional nerve trunks or via the bloodstream. The toxin produces increased muscle tone (rigidity) which results in muscle spasms or convulsions. The source of entry is commonly a wound made by an object contaminated with earth, manure or dust. In about one third of cases the wound may be no more than a superficial abrasion and be so small as to be undetected. Deep wounds heavily infected and containing foreign bodies, for example war wounds or injuries due to car accidents, provide the most favourable conditions for the germination of spores and the multiplication of bacilli.

The incidence of tetanus varies greatly in different parts of the world. Britain is a low-risk area, while developing countries, where standards of living are low and vaccination is unknown, have a relatively high incidence of infection. However, an increase in tetanus has been observed among narcotic addicts in the Western world. The incubation period is variable, ranging from a few days to three weeks, and this represents the interval between the infection of the wound, the production of toxin and its reaction with nervous tissue. The prognosis is regarded as poor if the incubation period is less than a week, while a long incubation period is often associated with a mild attack. *Cl. tetani* may remain in the tissues for months or even years until activated by irritation or trauma.

Oral manifestations

After a short period of non-specific prodromal symptoms the first manifestation of tetanus is tonic rigidity of the muscles of mastication. The patient presents with stiffness of the face followed by difficulty in chewing and swallowing. The edentulous patient may complain of an inability to insert his dentures. The spasm of the muscles of mastication often increases until the jaws are finally locked and the mouth cannot be opened. If the muscles of facial expression are involved, the corners of the mouth are drawn back, the lips protruded and the forehead is wrinkled giving the characteristic appearance of *risus sardonicus*. Other muscle groups become involved, for example the trunk and the proximal parts of the limbs. The spine may become arched (opisthotonos) and the chest fixed in a state of expiration. There is often a board-like rigidity of the abdominal muscles and, as the infection develops, the patient has difficulty in swallowing, and breathing is embarrassed because of restricted respiratory movements. Death occurs from exhaustion, aspiration pneumonia, or asphyxia due to respiratory muscle spasm. Rarely, tetanus has been described in relation to tooth extraction and oral sepsis.

Differential diagnosis
There are many causes of trismus which must be eliminated before a diagnosis of tetanus can be made, and the dentist rather than the physician may see the patient first. The more common causes of trismus are impacted third molar teeth with pericoronitis, periapical abscesses of the posterior teeth, tonsillitis, submasseteric abscess, retropharyngeal abscess, mumps, arthritis or dislocation of the temporomandibular joint, and fracture of the mandible.

Laboratory diagnosis

Laboratory investigations may not be helpful and the diagnosis of tetanus can usually be made from the clinical history and presentation. The treatment of tetanus must be started immediately a clinical diagnosis is made and not await microbiological results. The isolation of *Cl. tetani* from the wound is often difficult and uncertain due to the small numbers of organisms present and the heavy mixed growth of other microorganisms. If there is an open wound a swab of pus should be sent for culture and animal inoculation. A Gram-stained film should be examined for characteristic Gram-positive bacilli with round terminal spores and part of the sample should be cultured on a semiselective medium anaerobically at 37 °C. Typically the colonies of *Cl. tetani* are translucent and spreading. Identification is by a combination of biochemical tests and the demonstration of exotoxin.

Treatment

Treatment consists of the administration of antitoxin, protection of the airway, thorough wound débridement, antibiotics (usually penicillin), tetanus toxoid and sedatives to control muscle rigidity and spasm.

Further reading

Christie, A. B. (1981) *Infectious Diseases*, 3rd edn, Churchill Livingstone, Edinburgh

Dolby, A. E., Walker, D. M. and Matthews, N. (1981) *Introduction to Oral Immunology*, Edward Arnold, London

MacFarlane, T. W. and Samaranayake, L. P. (1989) Systemic infections. In *Oral Manifestations of Systemic Diseases*, 2nd edn, (eds J. H. Jones and D. K. Mason), W. B. Saunders, Philadelphia (in press)

Mims, C. A. (1982) *The Pathogenesis of Infectious Disease*, 2nd edn, Academic Press, London

Sleigh, J. D. and Timbury, M. C. (1986) *Notes on Medical Bacteriology*, Churchill Livingstone, Edinburgh

Tausig, M. J. (1984) *Processes in Pathology and Microbiology*, Blackwell Scientific, Oxford

Chapter 9
Fungal infections

A variety of diseases caused by fungi may appear in the oral and perioral regions, either as localized lesions or as manifestations of systemic mycoses. However, candida infections of the oral cavity are by far the predominant group of fungal infections which the dental practitioners in the Western world frequently encounter, although rare mycoses such as blastomycosis and mucormycosis may be seen, particularly in immunocompromised and sometimes in immigrant population groups. Hence, the emphasis of this chapter is on candida infections of the oral mucosa; the uncommon mycoses are briefly discussed at the end of the chapter (and see Table 9.5).

A number of local and general predisposing factors play a contributory role in the pathogenesis of oral fungal infections. Debilitated individuals, eg. those receiving antibiotics, steroid or cytotoxic therapy, are particularly predisposed, as are patients with uncontrolled diabetes mellitus, other hormonal disorders such as hypoadrenocorticism, and haematogenous and malignant disorders. Certain age groups, especially the very young and the very old, and the pregnant are prone to oral yeast infections.

Table 9.1 Factors that predispose the host to oral candidosis

Physiological:	Old age, infancy, pregnancy
Local trauma:	Mucosal irritation, denture wearing
Antibiotics:	Particularly broad-spectrum antibiotics
Corticosteroids:	Steroid inhalers, systemic steroids
Malnutrition:	High carbohydrate diet; deficiencies in iron, folate, vitamin B_{12}
Endocrine disorders:	Hypoendocrine states (e.g. hypothyroidism, Addison's disease), diabetes mellitus
Malignancies:	Blood disorders (e.g. acute leukaemia, agranulocytosis)
Immune defects:	AIDS, thymic aplasia
Xerostomia:	Due to irradiation, drug therapy, Sjögren's syndrome (primary and secondary), cytotoxic drug therapy

The effects of local factors are particularly well seen in chronic atrophic candidosis. Unclean dentures which may act as the reservoir of yeasts may be an aetiological cofactor, as is mechanical trauma due to an ill-fitting denture. Local use of antibiotics and corticosteroids may predispose to oral candidosis, together with other factors which are listed in Table 9.1.

Oral candidosis

Oral candidosis is a clinical entity which has been known since antiquity and manifests in a variety of ways. Accordingly, the disease has been categorized into several descriptive categories as localized candidoses which are confined to the oral and perioral tissues, and rare systemic disorders where candidosis may affect other organ systems as well as the oral cavity.

The variety of clinical types of candidosis is matched by a variety of causative yeast species. The imperfect fungus *Candida albicans* is the best known and the commonest but at least twelve other *Candida* species have been isolated from the oral

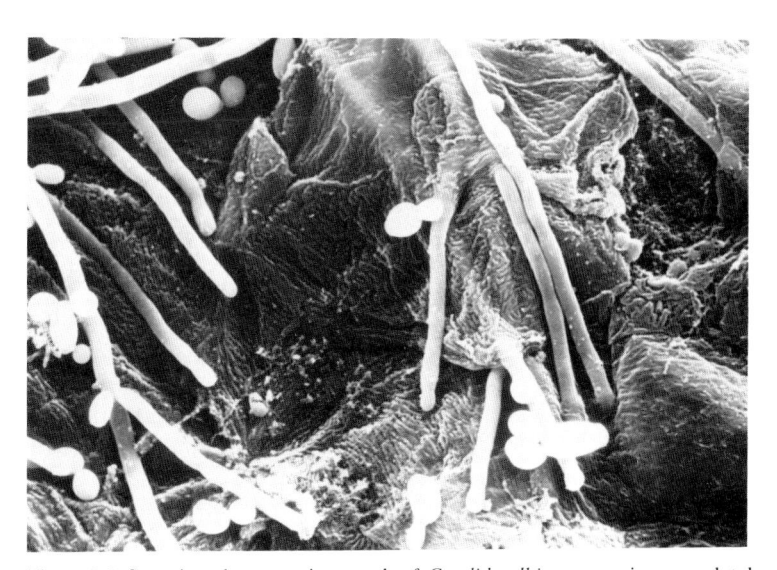

Figure 9.1 Scanning electron micrograph of *Candida albicans* growing on palatal epithelium of a rat showing clearly the blastospores and hyphal phases of the organism. Magnification ×3000, reduced to 60% on reproduction

cavity. They include *Candida glabrata, Candida tropicalis, Candida pseudotropicalis, Candida guilliermondii* and *Candida krusei*. In addition to *Candida* species other yeasts (*Rhodotorula* spp., *Saccharomyces* spp.) may also be present but make up only a small proportion of total yeast isolates.

Candida albicans is a dimorphic fungus and basically exists in the classical ovoid, yeast or blastospore phase (syn. blastoconidia) and hyphal or mycelial phases (Figure 9.1). They multiply primarily by the production of buds from ovoid yeast cells. Under adverse conditions they also form spores called chlamydospores which are not usually observed in clinical specimens but can assist in the laboratory identification of yeasts.

Oral carriage of *Candida* species in adults range from 20–60% and, on average, about 40% of humans harbour the yeast as an innocuous commensal. The yeast is also present in the throat, large bowel and vagina. The primary oral reservoir of the organism in carriers is the dorsum of the tongue.

Candidoses confined to the oral and perioral tissues

Candidoses which are relatively common but confined to the oral and perioral tissue can be classified as shown in Table 9.2.

Acute pseudomembranous candidosis (thrush)

Thrush, or acute pseudomembranous candidosis, is a well-known disease entity associated with a local disturbance or systemic illness. Thrush occurs in up to 5% of newborn infants and in some 10% of elderly, debilitated individuals. In the former it is the immature host defences which mediate the disease while in the latter it could be due to one or more of the predisposing factors mentioned in Table 9.1.

Thrush has recently acquired a renewed importance as it is one of the earliest and sometimes the initial manifestation of acquired immunodeficiency syndrome (AIDS). However, in association with AIDS this lesion may persist for months and the term 'acute' seems to be inaccurate in this context (see also Chapter 15).

The major clinical sign of thrush is the creamy-white, non-keratotic coating (pseudomembrane) found either on the mucosa of the tongue, soft palate, cheek, gingiva or pharynx. The coating is easily rubbed off to leave a red, raw and often bleeding base. The lesions vary in size from small discrete areas

Table 9.2 Classification of oral candidosis. Group I: Common disorders confined to the oral and perioral tissues

Subgroup	Acute/chronic	Disease state	Synonym(s)
1	Acute	Acute pseudomembranous	Thrush
2	Acute	Acute atrophic	Candida glossitis/glossodynia
3	Chronic	Chronic hyperplastic candidosis	Candida leukoplakia
4	Chronic	Chronic atrophic candidosis:	Denture sore mouth, denture-induced stomatitis
		Newton's Type I: pin-point lesions	Localized simple inflammation
		Newton's Type II: diffuse erythema	Generalized simple inflammation
		Newton's Type III: granular	Papillary hyperplasia
5	Acute/chronic	Angular cheilitis	Perlèche

to confluent white patches covering a wide area. Symptoms are usually slight and include such complaints as dryness or roughness of the mucosa. Patients do not normally complain of pain but if the lesion spreads to the pharynx or oesophagus it may cause a sore throat or dysphagia.

Histological examination shows great numbers of hyphae invading the superficial epithelial layers which are infiltrated by leucocytes. As a consequence the white coating is easily rubbed off or spontaneously shed as a pseudomembrane which consists of desquamated epithelium, keratin, fibrin, necrotic tissue and bacteria matted together by fungal hyphae.

Laboratory diagnosis
The lesion should be swabbed and sent to the laboratory together with a smear of the pseudomembrane (Table 9.3).

Table 9.3 Specimens required for the laboratory investigation or oral candidoses

Disease	Swab	Smear	Biopsy
Acute pseudomembranous candidosis	+	+	−
Acute atrophic candidosis	+	±	−
Chronic hyperplastic candidosis	+	±	+
Chronic atrophic candidosis:			
palate	+	+	−
denture	+	+	NA
Angular cheilitis	+	+	−

+, Useful; ± may be useful; −, inappropriate; NA, not applicable.

Management
Acute pseudomembranous candidosis can be treated by either polyene (nystatin or amphotericin B) or imidazole (miconazole) antifungal agents. The drugs should be administered topically either as suspensions, gels, creams or lozenges and resolution usually occurs within two weeks of treatment. Cases in which the disease persists should be investigated for previously unsuspected underlying disease.

Acute atrophic candidosis

This condition may arise in connection with prolonged steroid or broad-spectrum antibiotic therapy or a few days of topical antibiotics which can cause marked changes in the indigenous bacterial flora. The condition is less commonly called 'antibiotic sore tongue' or glossodynia, although it can affect any part of the oral mucosa, including the buccal mucosa and palate, in addition to the tongue. When the latter is involved there is marked depapillation and the affected areas generally become fiery-red, shiny and painful. It is perhaps the only form of candidosis which is consistently painful.

Although the condition is generally described in association with broad-spectrum antibiotic therapy either in local or systemic forms (e.g. tetracycline) it may present as a sequel of acute pseudomembranous candidosis where the membrane is shed leaving a red, atrophic mucosa. Indeed, closer examination may reveal small patches of thrush in protected niches such as the upper buccal sulcus.

As opposed to the pseudomembranous variety, histological examination reveals only a few blastospores and hyphae penetrating the surface layers of non-keratinized epithelium. An associated acute inflammatory infiltrate in the atrophic epithelium and the connective tissue is invariably present.

Laboratory diagnosis
Swabs and smears from the lesion may reveal the presence of yeasts in small to relatively large numbers in the affected areas (Table 9.3).

Management
It is often not possible to discontinue the antibiotic or steroid therapy which initiated the condition due to the serious nature of the underlying condition; therefore, the mainstay of treatment is long-term nystatin or amphotericin therapy (using the drugs in suspension). Patients should be advised to use antifungals regularly until a mycological cure of the condition is achieved (i.e. a negative result from laboratory investigations). Indeed, prophylactic antifungals may be necessary if the drug regime which precipitated the condition is to be continued.

Chronic hyperplastic candidosis (Candida leukoplakia)

Candida albicans can cause chronic lesions of the oral mucosa which persist as white patches (leukoplakia) and are indistinguishable from leukoplakias due to other causes. They usually occur as a single lesion on the mucosa, often on the inside surface of the cheek near the commissures or the surface of the tongue (Figure 9.2). The diagnosis can only be made with certainty by histological examination of biopsy specimens but

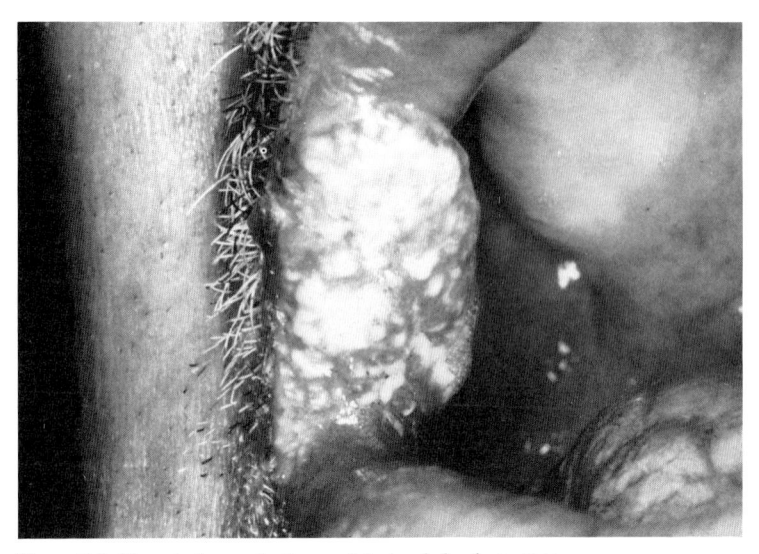

Figure 9.2 Chronic hyperplastic candidosis of cheek mucosa

other features of candidal infection such as angular cheilitis may be present.

The disease is typically seen in middle age although in the rare mucocutaneous candidosis syndromes (see below) it may be recognized at an early age as a manifestation of the syndrome. In contrast to the pseudomembranes of acute candidoses the plaques of hyperplastic candidosis can not be easily wiped off and are firm, whitish or speckled. Histologically the epithelium is parakeratinized and generally markedly hyperplastic. Candidal hyphae invade the parakeratinized layer more or less at right angles to the surface but never penetrate deeper than the glycogen-rich zone of the epithelium (Figure 9.3). A clinical feature of importance is the presence of red areas (speckled leukoplakia) and such lesions have a high propensity for malignant transformation.

It is not clear whether the candidal infection itself is the cause of leukoplakia or whether the lesions are primarily leukoplakic with a secondary candidal infestation. However, the resolution of at least some of the lesions by prolonged antifungal therapy adds credence to the former hypothesis.

Laboratory diagnosis
Biopsy is important as the condition may be premalignant,

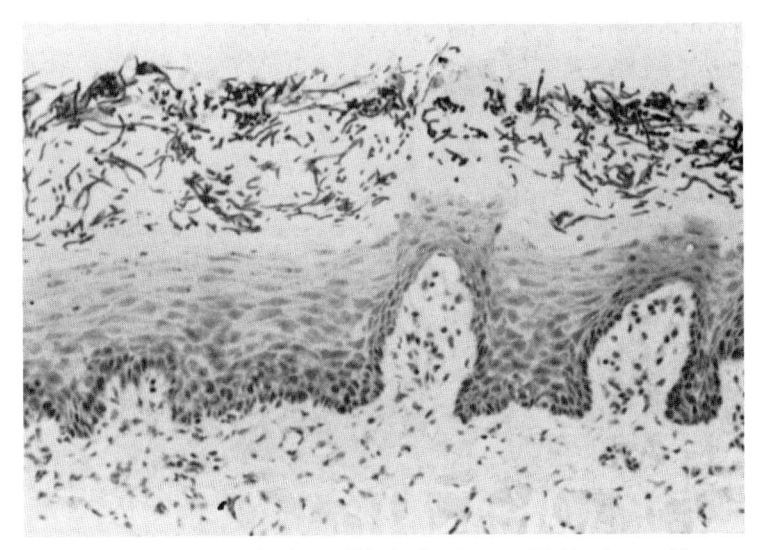

Figure 9.3 Chronic hyperplastic candidosis showing candidal hyphae and inflammatory cells infiltrating the superficial layers of the oral epithelium. Magnification ×250, reduced to 60% on reproduction

revealing varying degrees of dysplasia. In addition, swabs and smears from the lesion may yield positive or negative results (Table 9.3). Haematological and immunological investigations may be required to rule out the possibility of a rare mucocutaneous candidosis syndrome (see below).

Management
Long-term amphotericin (lozenges) or nystatin (pastilles) should be given four times a day until clinical resolution occurs. As antifungals alone may not resolve the condition, excessive hyperplastic tissue may require excision or cryotherapy, particularly when prostheses are required. Long-term review of patients is essential in view of the risk of malignant transformation.

Chronic atrophic candidosis (denture-induced stomatitis)

This is the commonest form of oral candidosis. It is prevalent in some 24–60% of denture wearers and may be associated with orthodontic appliances or an obturator. It is commonly found on the palatal mucosa beneath the fitting surface of the upper denture and both full and partial denture wearers are affected. Chronic atrophic candidosis is very rarely, if ever, found on the lower denture-bearing mucosa. The condition has been divided by Newton into three distinct clinical categories.

Type I Showing localized pin-point hyperaemia.
Type II The most common type of chronic atrophic candidosis. Presents with diffuse erythema and oedema of the denture-bearing areas of palatal mucosa. The affected area is bright red or dusky and sharply differentiated from the surrounding mucosa at the margins of the denture (Figure 9.4). The majority of patients do not usually complain of soreness and the condition is generally accompanied by angular cheilitis.
Type III If the type II condition is untreated for a prolonged period a hyperplastic epithelial reaction may occur resulting in a nodular type of lesion with intermittent atrophic areas.

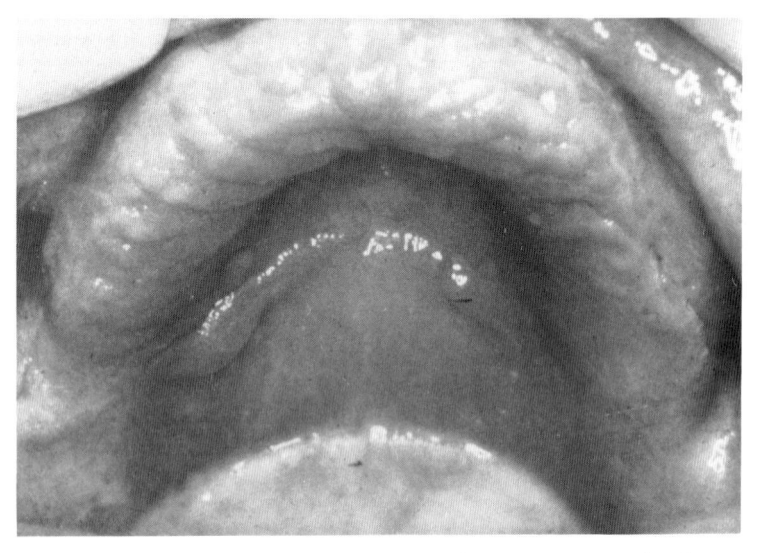

Figure 9.4 Chronic atrophic candidosis showing the erythematous and oedematous denture bearing mucosa

The vast majority of patients with chronic atrophic candidosis are otherwise healthy as local factors, such as trauma, poor denture hygiene and carbohydrate-rich diets, determine the pathogenesis. Occasional underlying factors are iron deficiency, xerostomia, folate deficiency and diabetes mellitus.

Laboratory diagnosis
Swabs and smears from the palate as well as the fitting surface of the denture are essential for laboratory confirmation of chronic atrophic candidosis (Table 9.3). Smears from the denture, taken with the help of a blunt edge of a plastic instrument, usually show numerous hyphae and blastospores of *Candida* together with a few leucocytes (Figure 13.4).

In refractory or recurrent forms of the disease haematological investigations for iron, folate and vitamin B_{12} may be required.

Management
As one of the factors involved in the initiation of the disease is continuous wearing of dentures which also act as reservoirs of yeasts, treatment involves denture hygiene measures and antifungal therapy. Patients should be discouraged from wearing dentures, particularly during sleep; the dentures should be soaked overnight in 1% hypochlorite or 2% chlorhexidine, both of which regimes have disadvantages as they may cause discoloration of the dentures, particularly the metallic components which tend to corrode when soaked in hypochlorite.

In addition, topical antifungal therapy with amphotericin (tablets or lozenges) or nystatin (pastilles) should be instituted. Similarly, ointments of these antifungals or miconazole gel should be applied regularly to the fitting surface of the prosthesis for 4 weeks. In Scandinavia, 0.2% chlorhexidine gluconate mouth-rinses have been successfully used to resolve the condition.

It is also important to treat any underlying disorder such as iron deficiency.

Angular cheilitis (angular stomatitis/perlèche)

This is most frequently seen in dental practice as a complication of chronic atrophic candidosis although it can be associated with any type of oral candidosis. The condition also occurs in dentulous patients of any age.

The disease has a multifactorial aetiology, one factor being infection by candida or staphylococci and, in some cases, a combination of these two organisms. The usual endogenous reservoirs of these organisms are the oral cavity and anterior nares, respectively. In a few instances β-haemolytic streptococci, particularly Lancefield group B, can also be demonstrated. Other factors involved in angular cheilitis are iron deficiency anaemia and vitamin B_{12} deficiency although cofactors, such as inadequate vertical dimension of dentures which results in deep folds at the angles, may be contributory. However, deep folds at the angles may occur in the elderly even in the absence of dentures as sagging facial tissues are an inevitable accompaniment of ageing. Salivary contamination and the residual moist conditions within the folds may lead to skin maceration and also promote fungal and bacterial growth which may either initiate or aggravate the cheilitis (Figure 9.5).

```
        Candida Species              Staphylococcus aureus
                        ↘          ↙
Iron, Vit B₁₂       →   Angular Cheilitis   ←   Skin creasing
deficiency                                       due to ageing
                        ↗          ↖
        Inadequate                   Systemic disorders
        dentures                     eg AIDS
```

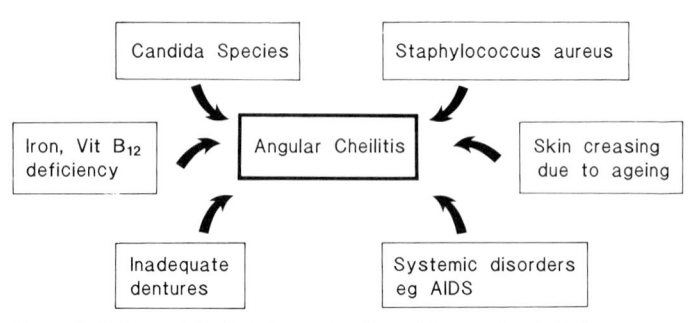

Figure 9.5 The aetiological factors implicated in angular cheilitis

Clinically the lesions vary from mere reddening at the angles to ulcerated and crusted fissures (Figure 9.6) and in a majority of cases a concomitant intraoral candidal infection is present. Some cases of staphylococcal angular cheilitis can be recognized clinically by a distinctive yellow crust, not unlike the typical lesion of impetigo. Since angular cheilitis is usually only a mild, if irritating, chronic infection, most patients will not seek medical or dental treatment.

Laboratory diagnosis
Swabs and smears from the angle(s) of the mouth should be sent to the laboratory. Since the lesions are often dry it is important that swabs moistened with sterile water or saline are used for such sampling.

Concomitant sampling for oral carriage of *Candida* species and nasal carriage of *Staphylococcus aureus* is useful to track down the reservoirs of infection. The anterior nares should be individually swabbed for staphylococci while the oral carriage of yeasts can be ascertained by an oral rinse sample (see Chapter 13).

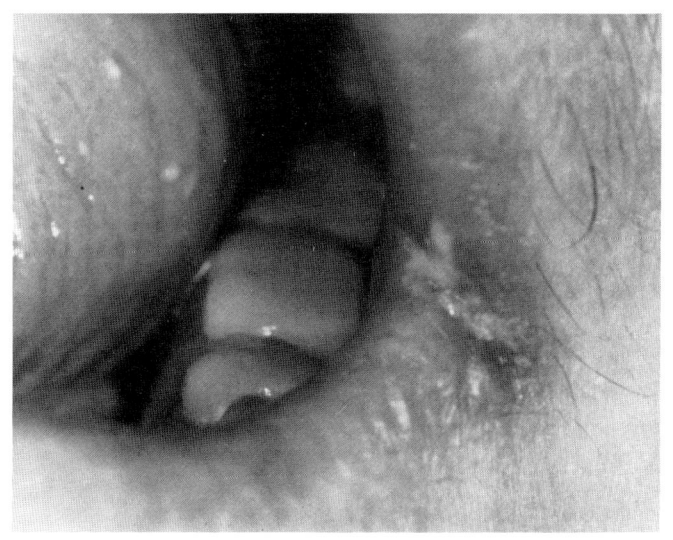

Figure 9.6 Angular cheilitis

Management
Elimination of the reservoirs of infection is central to the successful management of angular cheilitis. If intraoral candidosis is confirmed, appropriate treatment (see above) should be instituted. In addition, topical amphotericin (paste or ointment) or nystatin (ointment) should be applied four times a day for four weeks to both angles. Miconazole (gel) t.d.s. for 4 weeks is an alternative.

When *Staphylococcus aureus* is isolated from the angles then antibiotic sensitivity of the organism should be determined. If the organism is sensitive to fusidic acid (Fucidin) then this should be applied four times daily for 2–4 weeks. In the event of a nasal reservoir of infection, fusidic acid should be similarly applied to the anterior nares to eliminate nasal strains of staphylococci. Miconazole gel (an imidazole) can be used if the organism is resistant to fusidic acid as it has some Gram-positive bacteriostatic action.

Should resolution not occur then both patient compliance and underlying disease should be considered. Occasionally individuals are encountered with chronic nose–mouth habits which encourage the spread of staphylococci.

Due to the mild nature of angular cheilitis most patients will attend work normally and, depending on their occupation, the staphylococcal reservoir which they harbour could initiate outbreaks of disease, e.g. staphylococcal enteritis by food handlers and wound infections by nurses.

Miscellaneous oral candidoses

Two uncommon candidoses of the orofacial regions, which do not strictly fall into the classification given in Table 9.2, have recently been described. They are chronic oral multifocal candidosis and cheilocandidosis. Candida infection seems to play a major role in these diseases as they respond very well to antifungal therapy.

Chronic oral multifocal candidosis
This clinical entity is characterized by erythematous, plaque-like or nodular lesions in two or more of the following locations: commissural area, palate or dorsum of the tongue. The patients do not have any systemic disorder or other oral mucosal disease apart from chronic atrophic candidosis. The lesions resolve on prolonged antifungal therapy although palatal lesions in non-denture wearers, and nodular lesions particularly seen in the commissural area, have a high tendency to recur. Oral candidal lesions seen in AIDS patients are also multifocal in nature and somewhat similar to chronic oral multifocal candidosis.

Cheilocandidosis
Candida species are known to be the major aetiological agent in this disease entity which is characterized by ulcerative, granulating lesions of the vermillion border of the lower lip. The yeast would appear to invade previously traumatized tissue. The effect of sunlight as a possible initiating factor has been suggested.

Oral manifestations of mucocutaneous candidoses

Although candida infections of the mouth may occasionally spread to involve the adjacent skin, they are essentially localized infections that involve relatively small areas of the body surface. However, in a very small number of cases chronic candida infections are encountered that simultaneously affect mucous membranes of the mouth and vagina as well as several areas of skin and nails. The lesions may be few and mild or extensive and severe and are almost always chronic in nature. These lesions belong to the group of conditions designated as chronic mucocutaneous candidoses (CMC). These are classified in a variety of ways and subgroups 1 to 4 shown in Table 9.4 are examples of CMC which invariably present with oral lesions.

Familial chronic mucocutaneous candidosis

Transmission of this condition is probably autosomal recessive with persistent oral leukoplakic lesions starting in childhood. Lesions of the skin and nails develop later and are often associated with low serum iron levels.

Diffuse chronic mucocutaneous candidosis

This is a severe disease with susceptibility to fungal and bacterial infections, particularly of the upper respiratory tract. Cell mediated immunity is usually defective and iron deficiency anaemia may be present. Oral lesions are severe and may extend into the pharynx or larynx while the dermal lesions are grossly proliferative; therefore, when the face is affected it may lead to severe disfigurement and restriction in opening the mouth. Consequently dental treatment may be difficult.

Candida endocrinopathy syndrome

Candidal infections tend to be mild and dermal lesions infrequent in this disease which appears to be transmitted in an autosomal recessive form. The main features are the associated endocrine deficiencies, in particular hypoparathyroidism, although any other type of endocrine deficiency may develop (Table 9.4, subgroup 3). It is noteworthy that endocrine deficiency may appear many years after candidosis has appeared.

Candidosis in other immune deficiency states

In addition to CMC discussed above, persistent candidosis is a well-recognized complication of defective cell mediated immunity and is also seen in the following circumstances.

1. In severe combined immunodeficiency, Di George's syndrome and chronic granulomatous disease where oral candidosis is one of many complicating features consequential to natural immune deficiency (Table 9.4, subgroup 5).

Table 9.4 Classification of oral candidosis. Group II: Rare disorders (except subgroup 6) where candidosis remains superficial

Subgroup	Condition	Onset	Affected sites and clinical features
1	Familial chronic mucocutaneous candidosis	First decade of life	Mouth, nails, skin; mainly chronic hyperplastic lesions
2	Diffuse chronic mucocutaneous candidosis	Before 5 years of age	Mouth, nails, skin, eyes, pharynx and larynx; chronic hyperplastic lesions
3	Candidosis–endocrinopathy syndrome	By second decade	Mouth; associated with hypoparathyroidism, hypothyroidism, hypoadrenocorticism and diabetes mellitus; chronic hyperplastic lesions
4	Familial mucocutaneous candidosis	First year of life	Mouth, nails, scalp, flexures; chronic hyperplastic lesions
5a	Severe combined immunodeficiency	Childhood	Oral mucosa, skin, nails; pseudomembranous or hyperplastic lesions
5b	Di George's syndrome	Childhood	
5c	Chronic granulomatous disease	Childhood	
6	Acquired immunodeficiency syndrome	Any age	Oral mucosa, oesophagus; may manifest as pseudomembranous, erythematous or hyperplastic lesions

2. Thrush is one of the earliest premonitory signs of AIDS (Table 9.4, subgroup 6; see also Chapter 15).
3. Iatrogenic immunosuppression during organ transplantation can produce oral candidosis, particularly in the form of thrush.

Laboratory diagnosis
Chronic candidosis must be examined by taking swabs and smears from lesions and its histopathological nature confirmed by biopsy.

Appropriate clinical and laboratory investigations should be carried out to rule out immunological or endocrine dysfunction.

Management
As these are rare disorders, such patients are likely to be seen by staff in the hospital dental services. The mainstay of treatment is intravenous amphotericin or 5-fluorocytosine given alone or in combination. Newer imidazoles such as ketoconazole have been reported to be more effective and less toxic than the earlier drugs. Topical antifungals often lead to only temporary remission of the lesions.

Oral manifestations of systemic mycoses

As mentioned earlier, oral lesions have been described in a wide range of systemic mycoses. Fortunately, these are rare disorders which most dental practitioners in the Western world will not encounter.

Many of the diseases produce granulomatous lesions in the lungs and oral lesions are most likely to be seen in South American blastomycosis (paracoccidioidomycosis) and histoplasmosis. Further, the oral lesions are not clinically distinctive and diagnosis should be made by culture of the organism, biopsy, or both.

South American blastomycosis (paracoccidioidomycosis) is caused by *Paracoccidioides brasiliensis*, a yeast-like organism which also has a mould phase forming white mycelia. As the name implies, the organism is found only in the western hemisphere and a history of travel to, or residence in, that area can be an important diagnostic clue. The disease is characterized by granulomatous inflammation of skin, lymph nodes, lung and liver. Spreading ulcers of the tongue, gingivae and soft palate are common early findings.

Table 9.5 Uncommon systemic mycoses and their oral manifestations

Disease and representative agent	Main sites affected	Major oral manifestations	Frequency of oral affection
Aspergillosis *Aspergillus fumigatus*	Paranasal sinuses; rarely tongue, soft palate	Plaque formation and intense local pain in oral lesions	Uncommon
Blastomycosis (North American) *Blastomyces dermatitidis*	Tongue, oral mucosa, gingivae, lip, mandibular bone	Ulceration, sessile projections, indurated swellings, actinomycosis-like draining abcesses	Rare
Coccidioidomycosis *Coccidioides immitis*	Nasolabial folds, skin	'Verrucous granulomas' resembling carcinomas	Very rare
Cryptococcosis[a] *Cryptococcus neoformans*	Gingiva, hard and soft palate, mucosa, tonsillar pillar	Violaceous coloured nodules or granulations, swellings, ulcers	Uncommon
Geotrichosis *Geotrichum candidum*	Oral mucosa	Similar to acute pseudomembranes of candidosis	Uncommon
Histoplasmosis[b] *Histoplama capsulatum* *Histoplasma duboisii*	Oral mucosa, tongue, palate, gingiva, periapical area	Nodular, indurated or granular masses or tissue destruction with bone erosion	40%
Mucormycosis *Mucor* species	Extension from maxillary sinus through palate into the mouth	Sloughing ulcers with grey eschar and exposed bone (especillary maxilla); unilateral facial pain.	Common
Paracoccidioidomycosis (South American blastomycosis) *Paracoccidioides brasiliensis*	Hard and soft palate, gingiva, tongue	Papules or vesicles leading to ulcers; extensive local destruction	Common
Sporotrichosis *Sporotrichum schenckii*	Oral mucosa	Erythematous, ulcerative lesions leading to granuloma, or papillomas	Uncommon

[a] AIDS associated oral lesions have been described.
[b] May present to the dentist with initial oral manifestations.

Histoplasmosis is caused by the yeast *Histoplasma capsulatum* which has a characteristic ability to convert from the yeast to mycelial phase in response to temperature changes *in vitro*. Histoplasmosis is almost exclusively endemic in the United States, although the fungus has a worldwide distribution. The disease is characterized by pulmonary infection which may either be asymptomatic, acute or chronic. However, disseminated disease is frequently seen in the oral mucosa, gingivae and the tongue in the form of indurated ulcers or granular masses.

Oral features of systemic mycoses are summarized in Table 9.5.

Further reading

MacFarlane, T. W. and Helnarska, S. (1976) The microbiology of angular cheilitis. *British Dental Journal*, **140**, 403–406

MacFarlane, T. W. and Samaranayake, L. P. (1989) Systemic infections. In *Oral Manifestations of Systemic Disease*, 2nd edn, (eds J. H. Jones and D. K. Mason), W. B. Saunders, Philadelphia (in press)

Odds, F. C. (1988) *Candida and Candidosis*, 2nd edn, Baillière Tindall, London

Samaranayake, L. P. (1986) Nutritional factors and oral candidosis: a review. *Journal of Oral Pathology*, **15**, 61–65

Samaranayake, L. P. (1988) Oral candidosis: predisposing factors and pathogenesis. In *Dental Annual 1989* (ed. D. D. Derrick), John Wright, London

Samaranayake, L. P. and MacFarlane, T. W. (1989) *Oral Candidosis*, John Wright, London (in press)

Chapter 10
Viral infections

There are numerous species of viruses capable of infecting humans and the organisms are primarily classified according to the type of nucleic acid present in their genome, i.e. DNA or RNA viruses. However, other features such as the arrangement of the protein subunits (capsids) surrounding the nuclear material, the size of the virion and whether the virion is naked or enveloped are useful in subdividing and speciating the viruses (Table 10.1).

Table 10.1 Examples of virus classification according to structure and composition

Virus	Nucleic acid; configuration	Virus symmetry	Envelope	Approximate size (nm)
Herpesvirus	DNA ; DS	Icosahedral	Yes	150
Adenovirus	DNA ; DS	Icosahedral	No	70
Poxvirus	DNA ; DS	Complex	Yes	200
Parmyxovirus	RNA ; SS	Helical	Yes	200
Human immuno- deficiency virus	RNA ; SS	Delta- icosahedron	Yes	100

DS, double stranded. SS, single stranded.

Certain species of viruses show affinity for specific mammalian cells. This phenomenon, known as tissue tropism, is exemplified by herpes and hepatitis viruses which replicate essentially in epithelial cells and hepatocytes respectively. Viruses that show tropism for oral and facial skin epithelium and elicit clinical manifestations in these regions include herpes simplex virus (I and II), varicella zoster virus and the Coxsackie viruses. All of these viruses share a common feature by producing vesiculoulcerative lesions in the oral and perioral regions, usually accompanied by lymphadenopathy (Table 10.2). The papovavirus (e.g. papillomavirus) also infects oral epithelium but it produces papillary rather than vesiculobullous lesions. Cytomegalovirus and mumps virus demonstrate tropism for salivary gland tissue with sialadenopathy as the major clinical finding.

Table 10.2 Viral infections of the oral and perioral regions and their distribution sites

Virus/Disease	Distribution site
Herpes simplex I	
Primary herpes	
Gingivostomatitis	Marginal gingivae and mucosae with vesicles and ulcers
Skin	Localized dermal eruptions
Ocular	Palpebral and bulbar conjunctiva
Eczema hepeticum (Kaposi's varicelliform eruptions)	Generalized oral and dermal eruptions with pustule formation
Secondary herpes	
Labialis	Vesicles on vermillion border
Intraoral	Clusters of small vesicles and ulcers on palatal gingiva or mandibular gingiva
Varicella zoster	
Primary (chickenpox)	Trunk and face with a few isolated oral vesicles
Secondary (shingles)	Unilateral vesicles stopping at midline and following the distribution of branches of Vth cranial nerve
Coxsackie virus, group A	
Herpangina	Soft palate and faucial pillars
Hand, foot and mouth disease	Movable mucosa, arms below elbows, legs below knees
Paramyxoviruses	
Measles	Punctate microulcers near parotid duct orifice (Koplik's spots) 2–3 days prior to rash
Mumps	Unilateral or bilateral parotid swelling

Interaction of virus and host cells

When a virus comes into contact with a host cell one of a number of host–parasite interactions may occur. Firstly, the host cell may not be susceptible to virus attack, e.g. only hepatocytes become infected with hepatitis B virus. Secondly, the host cell may be susceptible and viral infection occurs. In the first stage of infection the virus becomes attached to receptors on the cell surface and in entering the cell it usually becomes covered with part of the cell membrane. Once inside the cell, the protein coat of the virus is removed and the viral nucleic acid released for the production of viral mRNA. It is by means of mRNA that a virus takes over the biosynthetic properties of the host cell. Virus mRNA then attaches to cell ribosomes and directs the synthesis of virus-specific proteins. Subsequently new virus particles are assembled within the host and finally released from the cell, either slowly by extrusion (budding) through the cell membrane, or rapidly as a result of cell rupture. The host cell then dies, producing a cytopathic effect (CPE). The third possibility is that the cell becomes infected but not killed. The host cell becomes transformed (as a result of viral nucleic acid becoming incorporated in its genome) to one which possesses malignant properties, i.e. unrestrained, proliferative growth. The fourth possible interaction is when the cell becomes infected and the virus remains within the cell in a potentially active state (latency) but produces no obvious effect on cell functions. The virus can become activated by a range of stimuli at a later time, producing tissue damage, e.g. recurrent herpetic infections and shingles.

Recovery and immunity to virus infection

The defence mechanisms of the host react to virus infection in both a non-specific and a specific fashion. The non-specific mechanisms include: (1) mechanical and chemical factors, e.g. the washing action of saliva, swallowing and subsequent inactivation of sensitive viruses in gastric acid; (2) the phagocytic and killing activity of polymorphonuclear leucocytes and mononuclear phagocytes; and (3) the action of interferon (probably the most important factor in recovery from acute viral infection) which is a complex of 'protein regulating molecules' which are released by virus infected cells. Interferon, a powerful inhibitor of virus replication, is then taken up by healthy cells which are rendered resistant to subsequent virus infection.

The specific defence mechanisms which are activated during viral infection are the humoral (antibody) and cell mediated (lymphokine production and T killer cells) responses. While antibodies are important in conferring long-term immunity to viral infection, their role in recovery from acute infection is uncertain, since patients with defects in their humoral immune response usually recover well from viral but not bacterial infections. Cell mediated immunity is important in recovery from acute infection, and patients with defects in cell mediated immunity often suffer from very severe bouts of common virus infections. A combination of (1) T lymphocytes which are cytotoxic for viral infected cells, and (2) lymphokine release which contains interferon and factors which attract, retain and activate macrophages at the site of infection, are involved in the cellular immune response.

Herpes virus infections

Herpes simplex I, herpes simplex II, varicella zoster, Epstein–Barr virus and cytomegalovirus belong to the herpes group of viruses. Oral and perioral infections related to these viruses are described below.

Primary herpes simplex infection

Primary herpetic stomatitis is caused by herpes simplex virus and is the commonest viral infection affecting the mouth. While type I is most commonly isolated, type II, which in the past was thought to be confined to the genitalia, has been increasingly found in herpetic stomatitis. The incubation period is about five days and the virus is transmitted by direct contact with recurrent skin lesions or infected saliva. Children may carry the virus asymptomatically or as convalescent carriers in saliva for several months but the virus is rarely isolated in adults after the primary lesions heal. If infection occurs in early childhood, the disease is usually subclinical or mild, being dismissed as 'teething', but if infection occurs in adults the symptoms are generally much more severe. In Western countries with a high standard of living, there is an increasing number of adult cases diagnosed, due to the absence of infection in childhood.

The infectivity of the virus is relatively low, although occasionally small outbreaks occur affecting children living closely together.

The clinical features of primary herpetic stomatitis consist of an initial stage where there is a mild to severe fever, enlarged lymph nodes and pain in the mouth and throat. Subsequently, variable numbers of vesicles develop haphazardly on the surface of the oral mucosa, although the tongue, buccal mucosa and gingiva are most commonly affected in primary gingivostomatitis. The vesicles quickly rupture to form small round or irregular superficial ulcers with erythematous haloes and greyish-yellow bases (Figure 10.1). The gingivae, if involved, are inflamed and bleed easily and the infection may be confused clinically with acute ulcerative gingivitis (AUG). In some patients, AUG may develop secondary to primary herpetic stomatitis. The mouth is usually painful and eating and swallowing may be very difficult, leading to hospitalization for a few days. The lesions are always self-limiting and heal without scarring within five to ten days. In some cases the lips may be grossly swollen with blood-stained crusting, simulating the appearance of erythema multiforme. Rarely a macular, papular rash may be associated with primary herpetic stomatitis. In immunologically impaired patients, herpes simplex type I virus produces chronic ulcerative stomatitis, which is characterized by large (1–1.5 cm), painful, shallow ulcers with a white base which develop on the soft palate, gingiva, buccal mucosa and tongue.

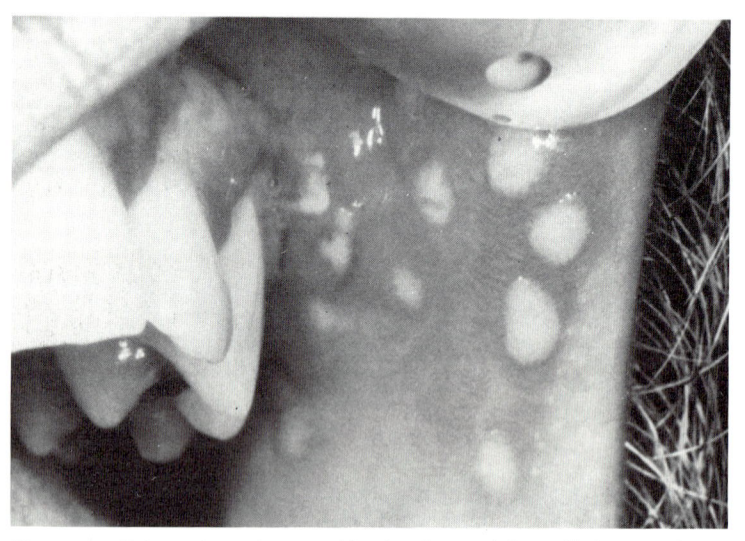

Figure 10.1 Primary herpetic stomatitis, showing mutiple small ulcers on the buccal mucosa. (From Partridge, 1987, *Dental Update* by permission)

Secondary herpes simplex infection

About 30% of patients who have had primary herpetic stomatitis develop recurrent infections later in life due to reactivation of the latent virus residing in the trigeminal ganglion. A stimulus such as acute sun exposure or trauma awakens the dormant virus which then travels along the neural axon and, after entering epithelial cells, multiplies and initiates the characteristic vesicle formation. The commonest lesion is that of herpes labialis or the 'cold sore'. The lesions develop at the mucocutaneous junction of the lip or on the skin adjacent to the nostril and are preceded by a premonitory prickling or burning sensation by approximately 24 h. Blisters then develop which enlarge, coalesce, rupture, become encrusted and then heal within 10–14 days (Figures 10.2 and 10.3).

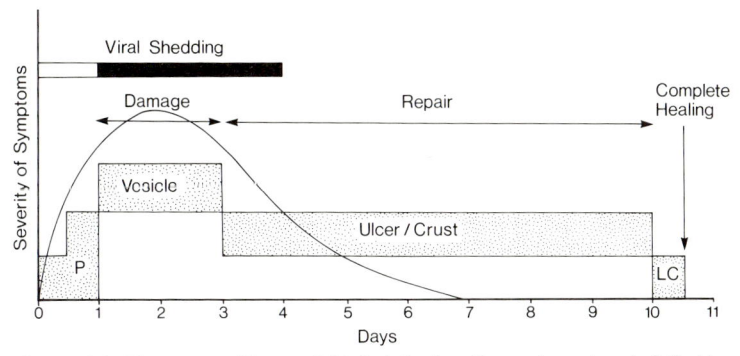

Figure 10.2 The stages of herpes labialis infection. P, prodromal period (itching, tingling and papule formation); LC, loss of crust

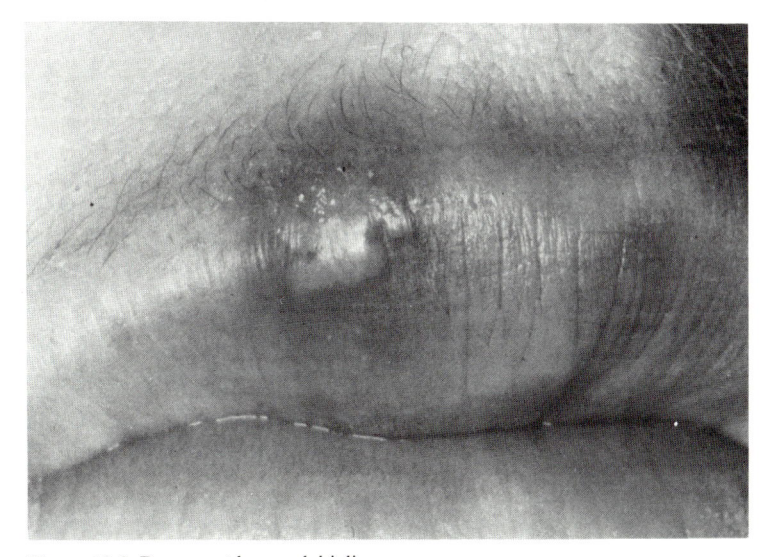

Figure 10.3 Recurrent herpes labialis

Intraoral recurrent herpetic infections have also been described and commonly involve the hard palate, alveolar ridges and gingiva. The stages of development of the intraoral lesions generally follow those described for lesions on the lip and appear as a cluster of small shallow ulcers with red irregular margins. Pain is not a common feature and the intraoral lesions may or may not recur intermittently for years.

Herpetic dermatitis/herpetic whitlow
Primary herpetic dermatitis is limited to a focal area on the skin and is characterized by pruritis, burning and pain. Multiple vesicles appear and persist for 4–5 days; they then burst and form crusting scabs which heal within 2–3 weeks. Dental practitioners who escaped exposure in childhood may contract herpetic dermatitis from patients having either primary or secondary herpes. Herpetic whitlow can also occur in dentists who have had a primary infection with herpes simplex. The commonest site of infection is the skin of the finger, resulting in an intensely painful herpetic whitlow (Figure 10.4). Herpetic whitlow may recur but less frequently than the perioral infection. Prompt diagnosis is essential to avoid exacerbation of the disease and transmission to patients.

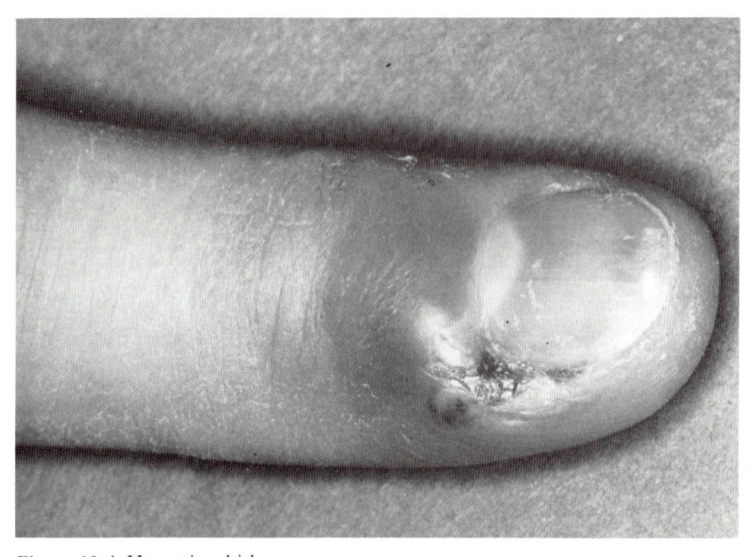

Figure 10.4 Herpetic whitlow

Ocular herpes
Primary herpetic infections can occur in the eyes causing dendritic ulcers in the cornea. Rarely, this may lead to blindness.

Eczema herpeticum
Children with atopic eczema or other chronic skin conditions (e.g. seborrhoeic dermatitis, pemphigus) may acquire an unusually severe form of primary cutaneous herpes known as eczema herpeticum or Kaposi's varicelliform eruptions. This condition is characterized by crops of oral and dermal vesicles which may become secondarily infected, leading to necrosis and sloughing of the epithelium.

Laboratory diagnosis (see also Chapter 14)
Direct examination   Smears should be stained with monoclonal fluorescent antisera to herpes simplex type I or II as this technique gives specific results and is useful in rapid diagnosis. Smears may also be examined with cytology stains to detect viral damage to cells. However a positive result is not specific for herpes simplex infection, and a negative result may mean that sampling was carried out too early or too late.

Culture   Herpes simplex is readily isolated in a variety of tissue culture systems from samples of oral lesions.

Serology   In primary infection, a four-fold or greater increase in antibody titre between the acute and convalescent sera is indicative of recent infection with herpes simplex. The demonstration of IgM antibodies by immunofluorescence techniques in a single sample can also be used in diagnosis. Cross-reactivity can occur between herpes simplex and varicella zoster antigens.

For a diagnosis of recurrent oral herpetic stomatitis to be made herpes simplex virus should be isolated from the lesion and a static antibody titre demonstrated.

Treatment
Moderate to severe cases of primary herpetic stomatitis should be treated with oral and topical acyclovir, together with other symptomatic measures. However, the use of acyclovir in recurrent herpetic infections should be limited to the immunocompromised and patients who have a past history of severe, extensive and/or frequently recurring lesions. The patient should apply the drug before vesicles form to obtain the best results.

Varicella zoster infection

Chickenpox and shingles are caused by the same virus, varicella zoster; chickenpox is the primary disease and shingles the reactivation of varicella virus present in latent form in sensory ganglia.

Chickenpox

Chickenpox is a common infectious disease which is present worldwide, and is usually contracted in childhood. In adults the disease tends to be more severe and sometimes complicated by viral pneumonia. Transmission of the disease is usually by direct contact or by droplet spread. The initial site of infection is the upper respiratory tract, and from there haematogenous spread occurs to other parts of the body. The incubation period of chickenpox is usually 15 to 18 days. Patients are infectious during the prodromal stage, especially if mouth lesions are present. Thereafter the infectivity falls and the patient may cease to be infectious once the skin lesions have become crusted.

The appearance of a rash is usually the first sign of the disease in children but in adults there may be a prodromal stage characterized by fever, headache, sore throat and general malaise. A scarlatiniform prodromal rash may occur which persists until the development of the true eruption.

The earliest stage of the skin eruption is pink maculopapules, which become vesicular and itchy within a few hours. The vesicles are easily ruptured and appear first on the back followed by the chest, abdomen, face and scalp, in that order. When the infection is fully developed, lesions are more profuse on the trunk than on the periphery of the body. Vesicles quickly become pustules, which dry to form a crust and heal within a few days. Chickenpox has a centripetal distribution, i.e. the greatest concentration of lesions being on the head and trunk, and all stages in the development of lesions are commonly seen together.

Oral manifestations
Before the typical skin rash develops, lesions may be found in the mouth, especially on the hard palate, pillars of the fauces and uvula, although any area of the oral mucosa may be involved. The oral lesions consist of small ulcers about 2 to 4 mm in diameter, surrounded by an area of erythema. The

vesicles are rarely found, due to their rapid rupture in the mouth. The lesions may be very painful in adults but probably less so in children, who rarely complain of discomfort while eating. Parotitis due to varicella has also been recorded.

Laboratory diagnosis (see Shingles below)
Treatment
Acylovir is not routinely prescribed for children or adults with chickenpox. However, systemic and, if necessary, topical acyclovir therapy is indicated in immunocompromised patients and in adult cases where viral pneumonia is complicating the infection.

Shingles (zoster)

Shingles is a localized eruption involving an area of skin affected by one or more sensory ganglia. The eruption consists of groups of vesicles on an erythematous base and is usually unilateral in distribution. Although the main age group for shingles is the elderly, the disease has been described for virtually all age groups. Local pain and paraesthesia commonly precede the eruption by several days and, in addition, some patients experience systemic upsets. Patients suffering from conditions which inhibit the immune response are particularly liable to severe attacks of zoster with a generalized varicella rash. While any sensory nerve may be affected, in over 50% of patients the thoracic ganglia are involved. The pain which precedes the rash is described as a burning sensation accompanied by intense hyperaesthesia. The rash consists of an erythematous area on which clusters of vesicles develop, and is confined to the segment supplied by the affected ganglion. The vesicles dry within a few days to form scabs which separate and heal, usually with little, if any, residual scarring. The communicability of zoster is far less than that of chickenpox and rarely infection may be transmitted by direct contact with infected vesicles. Since the virus of chickenpox and zoster are the same, such contact may produce chickenpox in susceptible individuals.

Oral manifestations
The trigeminal nerve is affected in about 15% of cases, with the ophthalmic, maxillary and mandibular divisions involved in that order of precedence. When the maxillary or mandibular divisions of the fifth cranial nerve are involved, the lesions of shingles may be found on the skin, on the oral mucosa or a

mixture of both (Figure 10.5). Severe localized oral pain often precedes the rash and in these circumstances an accurate diagnosis is very difficult since the patient may complain of toothache in the prodromal stage of infection. The most common intraoral sites are the anterior half of the tongue, the soft palate, and the cheek. The vesicles break down intraorally within a few hours to give very painful ulcerated areas with a yellowish-grey surface and erythematous borders. The oral lesions heal more quickly than the skin lesions and rarely scar. While the clinical presentation is usually unilateral, reports of bilateral infection have been reported.

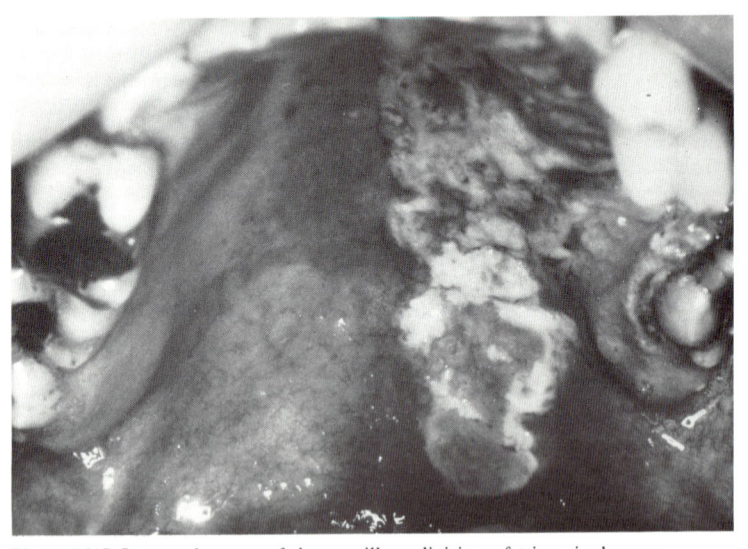

Figure 10.5 Intraoral zoster of the maxillary division of trigeminal nerve

In recent years herpes zoster infection of the mandibular and maxillary divisions of the trigeminal nerve have been implicated in tooth exfoliation and osteonecrosis of the jaws. The pathological changes are usually found in elderly patients, sometimes in association with malignant disease and treatment with cytotoxic drugs or irradiation. Rarely the condition has occurred in children. Varicella zoster has also been implicated in idiopathic root resorption. However, the precise role of varicella zoster in tooth loss is not clear, due to the complication of cytotoxic drugs and radiotherapy in a number of cases.

Laboratory diagnosis of chickenpox and shingles
The diagnosis of chickenpox and shingles is usually based on the case history and examination of the patient. In the case of patients where the clinical diagnosis is uncertain, the presence of the disease can be confirmed in the laboratory by sending: (1) vesicle fluid for electron microscopy and virus isolation; (2) smears from an ulcer for immunofluorescence; and (3) acute or convalescent sera to test for the presence of specific IgM antibodies to varicella zoster using an immunofluorescence technique.

Treatment
The best results with acylovir therapy in shingles are achieved when the drug is prescribed before vesicles form; patients then experience less pain and virus shedding is reduced. Once vesicles appear, acylovir is less effective, but should be given orally or, in severe cases systemically, due to the extreme pain experienced by patients, especially the elderly. On the other hand, the drug should always be prescribed when shingles develops in the immunocompromised.

Infectious mononucleosis

Whilst most of the viral diseases discussed in this chapter cause concomitant lymphadenitis, one of them produces a clinically pronounced lymphoid hyperplasia. The disease is infectious mononucleosis and the causative agent, the Epstein–Barr virus (a member of the herpes group of viruses).

Infectious mononucleosis is an acute infectious disease mainly involving children and young adults. In developed countries the majority of the patients are in the 15 to 25 age group but in underdeveloped countries children are most commonly infected. The source of the Epstein–Barr virus is the oropharyngeal secretions of patients suffering or convalescing from infectious mononucleosis; and the disease is transmitted by kissing. The virus has also been demonstrated in the oropharynx of healthy carriers. The incubation period of the disease is uncertain but may be as variable as five to thirty days.

Although infectious mononucleosis can be split into three clinical types, namely glandular, febrile and anginose, in many cases there is no sharp demarcation and the following features are common to all three: lymph node enlargement, fever and pharyngeal inflammation.

Oral manifestations
At the onset of the anginose type, the throat is painful and congested but exudate is absent. An enanthema consisting of clusters of fine petechial haemorrhages may be seen at the junction of the hard and soft palates. These lesions are not diagnostic as they are found in other virus infections of the respiratory tract. Later a white pseudomembrane develops on the tonsil. Exudative membranes, either localized or extensive, may develop on other parts of the oral mucosa and oral ulceration may be present. This form of the disease is common in patients between 15 and 30 years of age. A number of cases of infectious mononucleosis have been reported where the presenting signs and symptoms were unilateral pericoronitis with bilateral submandibular lymphadenitis and mild fever.

Laboratory diagnosis
The diagnosis of infectious mononucleosis may be possible on the typical clinical presentation of the patient but laboratory tests are required to confirm the diagnosis.

Haematology Venous blood (4 ml) should be placed in an EDTA container and a film and differential white blood cell count requested to demonstrate the lymphocytosis and atypical mononuclear cells (20%) which are present in the blood of patients with the infection.

Serology A rapid diagnosis can be made by testing an acute serum sample for the presence of IgM antibodies to the Epstein–Barr virus capsid antigen, using an immunofluorescence technique. A positive result can also be obtained by demonstrating a rising antibody titre using the Monospot or Paul–Bunnell tests.

Cytomegalovirus disease

Cytomegalovirus, a member of the herpes virus group, is relatively harmless and causes disease mainly in the immunosuppressed and in the newborn (see also Chapter 17).

Coxsackie A virus infections

There are two diseases caused by group A viruses which

produce oral signs and symptoms, namely hand, foot and mouth disease, caused mainly by Coxsackie A16 and, less commonly, by types A4, A5, A9 or A10, and herpangina, caused by Coxsackie A2, A4, A5, A6 and A8.

Hand, foot and mouth disease

General and oral manifestions
Hand, foot and mouth disease is a common viral infection of the mouth which occurs in small localized epidemics, usually among children although adults also become infected. It is easily diagnosed because of its classic distribution. The incubation period is about three to five days and usually resolution occurs within a week. A common, early symptom seen in school children is facial pain, combined with tenderness along the course of the parotid duct and the presence of a few small vesicles around the duct opening. The onset of the oral and skin eruptions is usually accompanied by headache, malaise and sore mouth and throat but in many cases there is little systemic upset. The oral lesions start as bright red macules 2–8 mm in diameter, which form thin-walled circular or oval grey vesicles with red areolae. As a result of trauma, vesicles are rarely seen, and the common presentation is shallow ulcers with a yellow base and a hyperaemic margin. Lesions are found on the pharynx, soft palate, gingivae, buccal mucosa and tongue. The palmar surface of the hands, the plantar surface of the feet and, in children under five years of age, the buttocks may be affected. The typical appearances of these skin lesions are usually bright-red macules, 3–4 mm in diameter, with pale translucent centres which develop into thin-walled bullae or small ulcers with a greyish surface surrounded by an area of erythema. Lesions in the mouth and on the hands or feet, or both, do not always occur and in some cases a mild fever with a few oral ulcers may be the only clinical findings.

Transmission of hand, foot and mouth disease from a patient to a dentist has been described but is a minor hazard.

Laboratory diagnosis
The virus can be isolated from saliva, from vesicle fluid or more easily from the faeces. A greater than four-fold rise in neutralizing antibodies is also detectable during the course of the illness. The disease, however, is relatively mild so that laboratory investigations may not be necessary.

Herpangina

General and oral manifestations
Herpangina is a systemic viral infection which is common in children although it may occur in any age group. The disease is characterized by sudden onset of fever, and sore throat which lasts for two to three days and is followed by the appearance of oral and pharyngeal lesions. The patient may experience anorexia, dysphagia, vomiting or abdominal pains. The disease is usually transmitted by droplets from infected individuals, by faecal contamination or by fomites.

The pharynx is often hyperaemic and the typical oral lesions associated with this disease develop on the mucosa of the anterior pillars of the fauces, on the hard and soft palate, tongue, uvula, pharyngeal wall and tonsils. The classical clinical presentation is small papulovesicular lesions about one or two millimetres in diameter with a greyish-white surface and surrounded by red areolae. The lesions may be multiple (up to about 12) or occur in isolated, small groups. The periodontium, tongue or buccal mucosa are very rarely affected and this assists in differentiating the infection from primary herpetic stomatitis. The disease lasts for about three to four days, then the fever abates and the oral lesions begin to heal. The infection may be very acute, with prostration of the patient, or he may experience no more than a mild pharyngitis. A rare complication of herpangina is acute parotitis and, when this occurs, difficulty may be experienced in differentiating the condition from mumps on purely clinical grounds and laboratory investigations are required.

Laboratory diagnosis
If a patient with herpangina presents with the classical signs and symptoms, the diagnosis may be made on clinical grounds. Where there is some doubt, a definitive diagnosis can be reached only after laboratory investigation.

A rapid diagnosis can be obtained by an immunofluorescence technique which measures IgM in acute sera. The best sample for virus culture is vesicle fluid but if this is not available a swab from an ulcer should be collected. Since Coxsackie A are enteroviruses, faeces should be cultured. In small outbreaks of these infections an acute and a convalescent sample of venous blood may be useful to demonstrate a rising titre of neutralizing antibody to the causative virus.

Paramyxovirus infections

Measles, mumps, parainfluenza and respiratory syncytial viruses are categorized as paramyxoviruses. Of these, measles and mumps are of concern in dentistry as they commonly manifest with oral signs or symptoms. Measles is discussed in detail below while mumps is described under salivary gland infections (Chapter 11).

Measles

Measles is a common childhood exanthem that can affect a significant proportion of the population if not successfully immunized. It is a highly infectious disease which occurs in small epidemics about every two years in developed countries, mainly affecting children between three and five years of age. Transmission of the disease is by droplet infection. Patients with measles release virus from their respiratory tracts during the prodromal period and for about two days following the appearance of the rash. In countries with a high standard of living the disease is mild, with a very low mortality rate, while in less affluent countries the mortality is high.

The clinical features of measles commence with a catarrhal stage during which the typical enanthemas of the infection are present. Initially the clinical features are those of a common cold, with conjunctivitis and a degree of photophobia. The buccal mucosa during this stage is erythematous and on the second day tiny, bluish-white, pinpoint spots surrounded by dark-red areolae appear on the buccal mucous membrane opposite the molar teeth. The number is quite variable but is usually five or six. These lesions are called Koplik's spots and in a few cases they may be found around the papillae of the parotid ducts. They are often very fleeting in appearance and do not usually remain for more than a day or two. When large numbers of lesions are present they may coalesce and involve considerable areas of the oral mucosa. The tongue is usually furred and when the furring is removed the enanthema is similar to the 'red strawberry tongue' which is found in scarlet fever.

The exanthema of measles appears on the third or fourth day after the catarrhal stage. The rash consists of fine, sparse, discrete macules, or maculopapules which vary in colour from pink to brick red. As the rash develops the Koplik's spots disappear. After two to three days the rash fades leaving a bluish-brown stain and occasionally fine desquamation of the skin. Recovery is rapid after this stage although a slight cough may persist for some weeks.

Gangrenous stomatitis and noma as a complication of measles are seen in certain sub-Saharan West African nations. In these patients, who are usually malnourished, healing of the ulcers associated with measles is delayed. Gangrenous lesions of the lip, cheek and maxilla may develop, and if the patient survives, scarring may produce oral stenosis requiring surgery to prevent starvation. In addition, herpes simplex virus has been reported in about half of the slowly healing, deep oral ulcers which follow measles.

Laboratory diagnosis

A diagnosis of measles is usually made on clinical grounds. When laboratory investigation is required, the presence of virus can be demonstrated within cells collected from Koplik's spots using direct fluorescent methods, or IgM antibodies detected using immunofluorescence.

Viruses and cancer

Many species of herpes viruses are directly associated with lymphoid and renal cancer in lower animals and certain oncogenic RNA viruses cause cancer in rodents. Three main groups of viruses have shown definitive links with cancer in man: Epstein–Barr virus in Burkitt's lymphoma and nasopharyngeal carcinoma; herpes simplex II in cervical cancer; and hepatitis B virus in hepatocellular carcinoma.

As far as the oral cavity is concerned, a number of epithelial neoplasms are thought to be associated with papillomaviruses which are epitheliotropic, DNA viruses. Some of these rare conditions include focal epithelial hyperplasia (Heck's disease), oral squamous papilloma, oral condyloma acuminata, oral verruca plana and verruca vulgaris and oral leukoplakia. Furthermore, a link between oral squamous cell carcinoma and human papilloma viruses has been suggested by some workers.

Further reading

Evans, A. S. (1976) *Viral Infections of Humans. Epidemiology and Control*, Wiley, London

Hooks, J. J. and Jordan, G. W. (1982) *Viral Infections in Oral Medicine*,

Elsevier–North Holland, Amsterdam
MacFarlane, T. W. and Samaranayake, L. P. (1989) Systemic infections. In *Oral Manifestations of Systemic Diseases*, 2nd edn, (eds J. H. Jones and D. K. Mason), W. B. Saunders, Philadelphia (in press)
Mims, C. A. and White, D. O. (1984) *Viral Pathogenesis and Immunology*, Blackwell Scientific, Oxford
Sommerville, R. G. (1983) *Essential Clinical Virology*, Blackwell Scientific, Oxford
Timbury, M. C. (1983) *Notes on Medical Virology*, Churchill Livingstone, Edinburgh

Chapter 11
Salivary gland infections

Salivary gland infections (bacterial or viral sialadenitis) are not uncommon. The most frequently encountered infection of the salivary glands is caused by the mumps virus. However, a number of other agents, particularly bacteria, are responsible for acute sialadenitis.

The parotid glands are more commonly infected than the submandibular salivary glands and consequently most reports deal with bacterial parotitis, although the submandibular glands may be similarly affected. Infections of the accessory salivary gland can occur but are very rare (Table 11.1). Apart from mumps, the majority of salivary gland infections are seen in adults. However, there are two rather ill-defined clinical entities termed neonatal suppurative parotitis and recurrent parotitis of childhood (chronic juvenile parotitis) which are confined to the first decade of life.

Table 11.1 Classification of salivary gland infections

Type of infection	Gland usually affected	Predisposing factor(s)
Mumps (endemic parotitis)	Parotid	No prior exposure to virus
Acute suppurative parotitis	Parotid	Severe xerostomia (e.g. Sjögren's syndrome), localized and diffuse abnormalities of the salivary glands
Obstructive sialadenitis	Submandibular	Sialolith, foreign bodies, ductal strictures, mucus plugs
Suppurative and chronic recurrent parotitis of childhood	Parotid	Congenital or acquired abnormality of ductal system
Rare miscellaneous disorders, e.g. tuberculosis, actinomycosis and fungal infections	Parotid or submandibular	Systemic infection by specific agents, e.g. *M. tuberculosis*

As in many other infectious diseases, initiation and progression of salivary gland infections would appear to depend on the host resistance to infection and the virulence of the causative organism(s). The decrease in host resistance in the present context could be either due to general factors such as debility and dehydration or due to local factors including sialolithiasis, ductal strictures and salivary gland pathology. Factors important in salivary gland infections are shown in Figure 11.1.

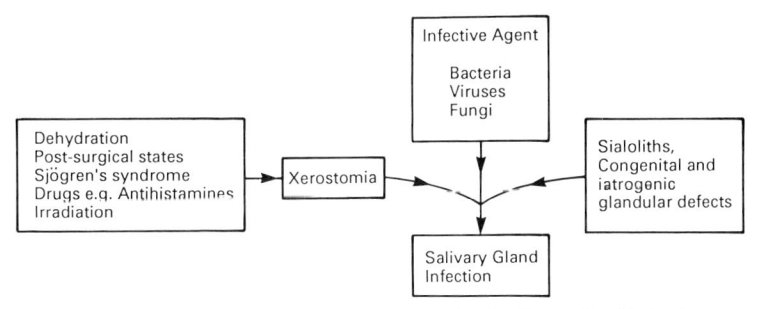

Figure 11.1 Factors important in the pathogenesis of salivary gland infections

Viral infections of salivary glands

Mumps (endemic parotitis)

Aetiology and epidemiology
Mumps is caused by an RNA paramyxovirus and is characterized by chronic inflammation and enlargement of salivary glands. It is frequently seen in winter and spring, and clinical or subclinical infection may occur at all ages but is most common in childhood.

The incubation period of mumps is usually 16–21 days but one month is not unusual. The saliva of patients who are incubating mumps is infectious for a few days before parotitis develops and up to two weeks after the onset of clinical symptoms. Mumps is transmitted by direct contact with saliva and by droplet spread.

Clinical presentation
There is a prodromal period of pyrexia, sore throat, furred tongue and usually a complaint of pain on chewing. Two early signs of mumps sometimes seen are the reddening of the opening of the parotid duct, and pain or tenderness on upward

pressure beneath the angle of the lower jaw. As a result of infection, the salivary glands increase in size, the degree of swelling varying from little visible enlargement to gross swelling. The glands may be firmer than normal or very hard. Usually both parotid glands become enlarged at the same time but in a number of cases a single gland may be affected or there may be a delay of up to five days between the first and second glands showing evidence of infection. As a result of inflammation and oedema, the salivary flow rate is low and patients may have a non-specific stomatitis and halitosis. Pain may be acute during salivation and parotid involvement may lead to severe trismus and earache. Salivary glands other than the parotid may become enlarged and, in the case of the submandibular glands, the swelling usually occurs a few days after the parotids and may reach their peak while the parotid glands are decreasing in size. In some 10% of cases, enlargement of submandibular glands occurs without parotid involvement.

There is wide variation in the clinical course of mumps, ranging from a mild upset lasting a day or two to a severe illness with high fever lasting 10–14 days, but complete recovery is usual.

The complications of mumps are simply other manifestations of the disease less common than sialadenitis. These usually follow parotitis but may precede it or occur without salivary gland involvement. Meningoencephalitis (30%) and orchitis (25% of adult males) are common, while rare complications include oophoritis, thryroiditis, neuritis, myocarditis and nephritis.

Laboratory diagnosis
The diagnosis is normally made on clinical grounds. In patients who have an unusual clinical presentation, laboratory investigations are required. Saliva, preferably pure parotid saliva collected by cannulation, may be examined by electron microscopy for typical virus particles, or inoculated into tissue culture. These techniques are not carried out routinely in all virus laboratories and the usual method of diagnosis is the demonstration of antibodies to mumps virus antigens using serological tests, e.g. the detection of IgM antibodies using immunofluorescence.

Other viral infections

Although mumps is the commonest viral cause of sialadenitis, cytomegalovirus, a member of the herpes group, can also infect salivary glands and cause clinical disease. The disease called cytomegalic inclusion disease (salivary gland inclusion disease) affects newborns, children and adults and has multiple systemic manifestations, including salivary gland enlargement. The disease is so called because of the large, doubly-contoured 'owl eye' inclusion bodies within the nucleus or cytoplasm of duct cells of the parotid gland.

Other viruses which have been implicated in non-suppurative sialadenitis include parainfluenza types 2 and 3, echo and Coxsackie viruses. These, however, are very rare disorders.

Bacterial infections of salivary glands

Acute suppurative parotitis (bacterial sialadenitis)

In the past this infection was primarily a disease of dehydrated or postoperative patients, particularly following gastrointestinal surgery but, with the introduction of fluid balance and antibiotic prophylaxis, suppurative parotitis is extremely rare in these groups. The infection is now usually related to adult patients with salivary gland abnormalities.

Aetiology
In health, potentially pathogenic bacteria in the mouth are prevented from ascending the duct and invading the salivary gland tissue by the flushing action of saliva. However, if the flow of saliva is greatly reduced or stopped, e.g. in patients who become dehydrated after gastrointestinal surgery or patients with Sjögren's syndrome (a progressive degenerative disease affecting salivary gland tissue), a retrograde infection via the salivary duct may ensue. Predisposing factors include: (1) drugs which reduce salivary flow, e.g. diuretics, certain antihistamines, tranquillizers and anticholinergics; (2) localized salivary gland abnormalities (calculus, mucus plug or benign strictures); and (3) generalized sialectasis, which is diagnosed radiographically by circumductal leakage of contrast medium previously introduced into the ductal system (Figure 11.2).

Clinical presentation
The clinical presentation is variable; unilateral, or sometimes bilateral, swelling of the parotid gland may be present and may last for days or weeks. Swelling may be limited to the parotid gland or, in more severe infections, extend locally, involving the pre- and postauricular area. The ear lobes may be displaced

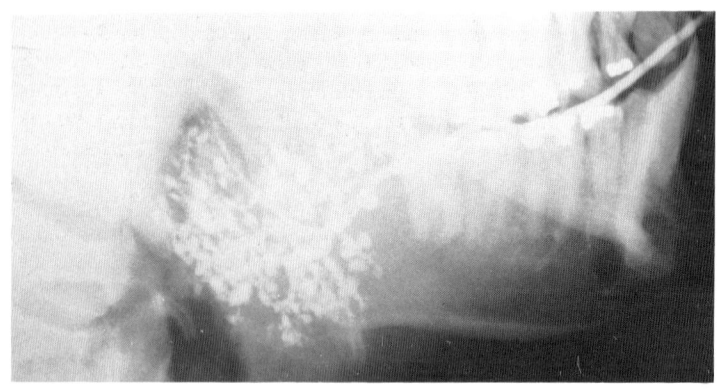

Figure 11.2 A sialogram of the parotid showing sialectasis which predisposes the gland to infection

laterally and purulent salivary secretions seen at the duct orifice. Pain and swelling may limit opening of the mouth. Usually there are no systemic symptoms but occasionally patients have a raised temperature, experience chills and have a leucocytosis. The patient suffers bouts of acute exacerbation of infection followed by periods of remission, the net result being replacement fibrosis of the parotid gland.

The main clinical features which distinguish mumps from bacterial parotitis are the absence of inflammation of the overlying mucosa and the absence of frank pus when milking the salivary duct.

Collection of specimens
Because of the variety of bacteria which may be involved, it is essential that a sample of pus is collected before antibiotics are prescribed. The details of the method for sampling parotid pus and transporting the samples to the laboratory are described in Chapter 13. In brief, wherever possible pus should be aspirated through a small-bore catheter attached to a syringe, or collected aseptically on a cotton wool swab and sent immediately to the laboratory. If the sample is believed to be contaminated with saliva or mucosal flora, this information must be recorded on the request form. If a sample of pus cannot be obtained from the duct and if the lesion is reasonably localized, aspiration through the skin may be attempted.

Microbiology
The microbiology of acute suppurative parotitis is poorly docu-

mented, most studies being retrospective and using methods of sample collection which were likely to produce a pus sample contaminated with saliva. The microorganisms which have been isolated from bacterial salivary gland infections are shown in Table 11.2, and both mono- and polymicrobial infections have been described.

Table 11.2 Bacteria commonly isolated from bacterial parotitis

Common isolates[*]
α-Haemolytic streptococci
Staphylococcus aureus

Less common isolates
Haemophilus spp.
Eikenella corrodens
Bacteroides spp.
Anaerobic streptococci

Rare isolates
Neisseria gonorrhoeae
Mycobacterium tuberculosis
Actinomyces spp.
Treponema pallidum

[*] Polymicrobial infections with two or more bacteria are common.

From the examination of the literature it would appear that α-haemolytic streptococci and *Staph. aureus* are the most common aetiological agents of acute suppurative parotitis. However, the frequency of isolation of *Staph. aureus* is gradually diminishing, particularly because bacterial sialadenitis is now uncommon in hospitalized patients for reasons mentioned earlier. In a recent study of 27 non-hospitalized patients with acute parotitis, 50% isolates were found to be streptococci as opposed to a mere 3% of *Staph. aureus* isolates. Other bacteria isolated included *Haemophilus* spp., *Eikenella corrodens*, *Bacteroides* spp. and anaerobic streptococci.

Management
The treatment of choice is parenteral antibiotic therapy guided by culture of pus and sensitivity tests (see Chapter 13). In the interim, however, a penicillinase-resistant penicillin, such as flucloxacillin, should be administered. Sometimes a Gram-stained smear of the pus is useful in deciding initial antibiotic therapy, as shown in Table 11.3. Instillation of antibiotics via

the salivary duct is preferred by a few workers. Oral hygiene is extremely important and salivation should be encouraged by increased fluid intake and the use of sialogogues. Surgical drainage of the affected glands may be required in severe cases.

Table 11.3 Use of Gram film in deciding initial antibiotic therapy in patients with acute suppurative parotitis

Appearance in Gram film	Likely causative agent	Antibiotic therapy
Gram-positive cocci in pairs and small clusters	*Staph. aureus*	Flucloxacillin
Gram-positive cocci in pairs and short chains	Streptococcal infection	Amoxycillin or erythromycin
Small Gram-negative pleomorphic rods	*Haemophilus* spp. *Bacteroides* spp.	Amoxycillin

Once the acute condition has resolved, the patient should be referred for sialographic investigation of the affected gland(s). This procedure is important to identify correctable salivary gland abnormalities, such as the presence of mucus plugs, benign strictures and calculi, which may lead to recurrence of further attacks. Sialography should never be attempted during the acute phase of the illness.

Sequelae
If acute bacterial parotitis is untreated, severe complications may ensue, especially in debilitated patients. Due to the thick connective tissue capsule around the parotid gland, spontaneous drainage of pus is uncommon with only small amounts being lost via the duct; therefore, inflammation and oedema extend into the neck and may cause respiratory obstruction, cellulitis, septicaemia or osteomyelitis of adjacent facial bones. Mortality as a result of septicaemia is a real threat (up to 20%) in elderly, debilitated patients.

Recurrent parotitis of childhood

This disease is of uncertain incidence but is believed to be much less common than acute suppurative parotitis in adults. It can affect children from 3 months of age and usually responds to empirical antibiotic therapy. The microbiology of this condition is poorly documented, although a few studies have suggested that *Staph. aureus* and/or enteric Gram-negative bacilli are usually present. A number of predisposing factors have also been suggested, e.g. abnormalities of the ductal system due to congenital factors or secondary to mumps, foreign bodies in the parotid duct, and trauma from orthodontic appliances.

Treatment consists of removal of any recognized aetiological agents, short- or long-term antibiotic therapy, ductal lavage and surgical drainage, depending on the clinical situation. As spontaneous resolution of the disease occurs during puberty, radical surgical therapeutic measures are contraindicated.

A note on chronic parotitis

Some adults are affected by recurrent bouts of parotitis. This is mainly due to the persistence of the aetiological agent (e.g. sialolith) which initiated the condition and subsequent low-grade ascending infection. Some clinicians prefer to call this condition chronic parotitis (or acute exacerbation of chronic parotitis), rather than acute parotitis, as the patient presents with the infection at different intervals over a prolonged period of time.

Submandibular sialadenitis

Submandibular sialadenitis is less common than acute suppurative bacterial parotitis. However, submandibular (submaxillary) glands may be affected with any of the problems already discussed or infected by any of the previously mentioned organisms.

Most bacterial infections in the submandibular glands are associated with obstructive ductal disease such as the presence of sialoliths and strictures. The management of submandibular sialadenitis is therefore dependent upon removal of ductal obstructions, duct dilatation or surgical revision of ducts. Nevertheless, aforementioned regimes such as collection of pus, culture and sensitivity studies, appropriate antibiotic therapy, rehydration and oral hygiene are necessary adjuncts.

Rare miscellaneous bacterial infections of salivary glands

Salivary gland infections by organisms such as *T. pallidum* and *N. gonorrhoeae* have been rarely documented in the literature. These diseases are characterized by unilateral or bilateral involvement and painless swellings of the gland. Similarly actinomycosis and tuberculosis of salivary glands have been reported and the routes of spread are thought to be (1) ascending infection via the ducts from the oral cavity, and (2) contiguous spread from foci of infection subjacent to the salivary glands (see Table 11.1). In addition, reactivation of tubercle bacilli deposited years previously by haematogenous or lymphatic spread from respiratory tuberculosis has been proposed. Clinical presentation of salivary gland tuberculosis takes two main forms: a chronic encapsulated type which takes months or years to develop and an acute type which develops within a few weeks or days. Rarely, facial paralysis may be a presenting symptom.

Oral infection consequential to xerostomia

Reduction or absence of salivary flow may lead to significant changes in the oral ecosystem, thereby affecting the rate and progression of diseases such as caries, periodontal disease and mucosal candidosis. For a fuller description of effects of xerostomia on oral infections see Chapter 12.

Further reading

Greenberg, M. S. (1984) Salivary gland disease. In *Burket's Oral Medicine* (ed. M. A. Lynch), Lippincott, Philadelphia, pp. 432–50

Lamey, P-J., Boyle, M. A., MacFarlane, T. W. and Samaranayake, L. P. (1987) Acute suppurative parotitis in out-patients: microbiological and post-treatment sialographic findings. *Oral Surgery, Oral Medicine and Oral Pathology*, **63**, 37–41

Mason, D. K., and Chisholm, D. M. (1975) *Salivary Glands in Health and Disease*, Saunders, London

Seifert, G., Miehlke, A., Haubrich, J., and Chilla, C. (1986) *Diseases of the Salivary Glands*, Thieme, Stuttgart

Chapter 12
Infections in medically compromised patients

From a functional point of view, a compromised host is any patient whose normal defence mechanisms are impaired, making the individual more susceptible to infection. There is a wide variety of conditions and illnesses that suppress normal defences and allow infections to develop. These conditions may range from some of the more common entities, including damaged heart valves and diabetes, to immunodeficiency diseases which are relatively uncommonly encountered in the dental practice. Frequently, infections seen in compromised individuals are due to true exogenous pathogens which also cause disease in non-compromised hosts. However, a significant proportion of infections in compromised patients is caused by both endogenous and exogenous organisms with either a high or low degree of virulence and pathogenicity. These infections, referred to as opportunistic infections, are often unusually severe and may present to the clinician with atypical manifestations.

Advances in medical and surgical treatment over the past few decades have produced an increase in the number of patients presenting for dental treatment who are compromised in a number of ways. Therefore, the main aim of this chapter is to concentrate on the infection risks experienced by such patients and to indicate why it is important for dentists to have an understanding of the problems involved in their management. The first part of this chapter deals in some detail with the dental management of the infective endocarditis patient, while the remainder addresses the treatment of oral infections in patients with immunodeficiency diseases and xerostomia.

Infective endocarditis

As transient bacteraemias caused by dental and related procedures are known to play a major role in the pathogenesis of infective endocarditis, all dental practitioners should possess a sound knowledge of the disease process and the rationale of preventive therapy.

Although infective endocarditis can rarely develop on a normal heart valve, an abnormal valve is usually involved in the pathogenesis of the disease. The main causes of cardiac valvular disease which predispose to infective endocarditis are shown in Table 12.1.

Pathogenesis

The development of infective endocarditis entails the sequential interaction of several events as shown in Figure 12.1. A breach of the integrity of the endocardium, or an abnormality of the endocardial surface *per se*, is the initial pathological event which makes the valvular surface eventually succumb to infection. Such a breach may occur due to the acute inflammatory valvulitis of rheumatic fever (consequential to *Streptococcus pyogenes* infection). In congenital heart diseases such as aortic valve disease and ventricular septal defect (Table 12.1) alterations of the blood flow patterns (haemodynamic turbulence) can result in a tendency for the deposition of fibrin and platelets at foci where high velocity jets of blood hit the valvular surface (Figure 12.1). These microscopic platelet aggregations may detach and emobolize harmlessly or stabilize and consolidate through fibrin deposition. The result is the formation of a

Table 12.1 Cardiac valvular disease which predisposes to infective endocarditis

Disease	Degree of risk
Aortic valvular disease Prosthetic valves Mitral insufficiency Ventricular septal defect Patent ductus arteriosus Coarctation of aorta Previous infective endocarditis	High
Mitral valve prolapse and stenosis Pulmonary and tricuspid valve disease Degenerative (calcific) aortic valve disease Nonvalvular intracardiac prosthetic implants	Intermediate
Atrial septal defect Coronary artery disese Cardiac pacemakers Arteriosclerotic plaques	Low/negligible

thrombus at the site which is thus a potential trap for circulating microbes. Such sterile thrombus formation is sometimes referred to as non-bacterial thrombotic endocarditis. This tendency towards platelet aggregation is not only limited to areas of damaged or denuded endocardium; platelets also have the potential to adhere to other 'foreign' surfaces such as prosthetic valves. It is important to realize that non-bacterial thrombotic endocarditis is usually asymptomatic, provided impaction of emboli in vital areas does not occur.

The next crucial event in the pathogenesis of the disease occurs when microorganisms circulating in the blood attach to or become trapped in the thrombotic endocardium or the prosthetic device. The resultant platelet–fibrin–bacterial mass, now

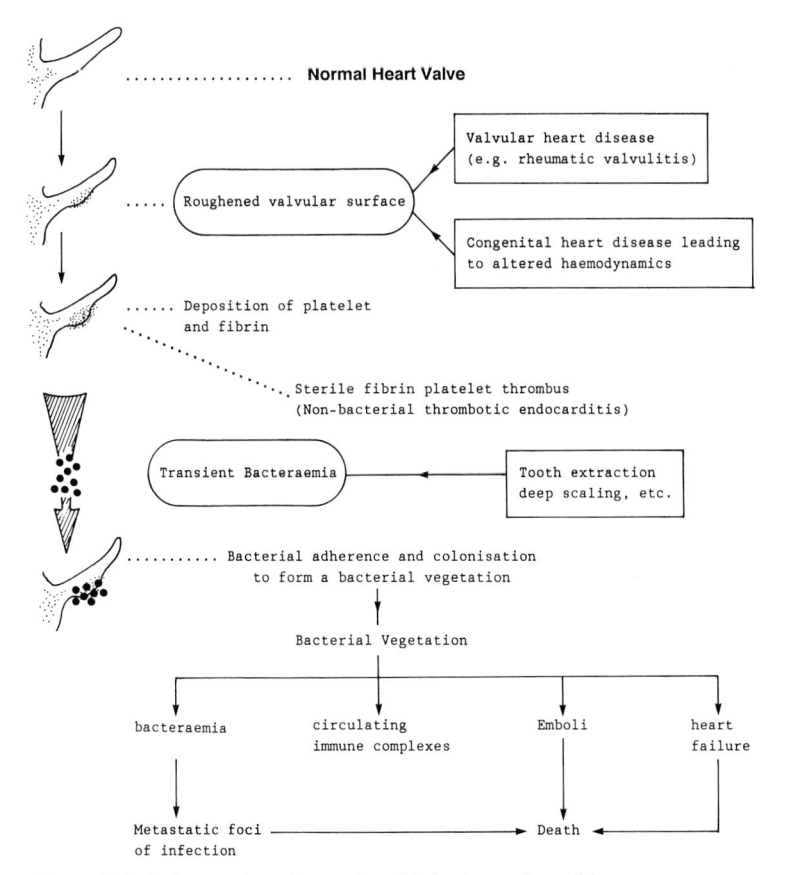

Figure 12.1 Pathogenesis and sequelae of infective endocarditis

170 Infections in medically compromised patients

called the bacterial vegetation, constitutes the primary pathology of infective endocarditis (Figure 12.2). Although the detailed mechanics of the microbial ecosystem of the vegetation is not clear, it is probable that once the organisms are attached to the lesion they multiply and colonize this niche in an exuberant manner. As a result, further aggregation of platelets and fibrin deposition would ensue, thus protecting the organisms from the body defences. The organisms now reside in a sanctuary inaccessible to phagocytes by virtue of the fibrin–platelet barrier. Further, the bacteria may be sheltered from antibiotics and host antibodies, as the vegetation is essentially avascular in nature. As a result, it is necessary to use an intensive course of prolonged, high-dose antibiotic therapy to eradicate such an infective focus.

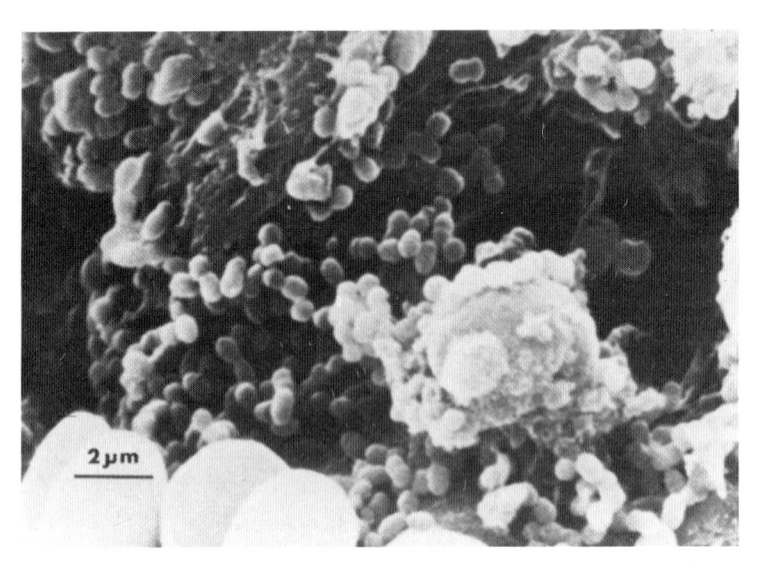

Figure 12.2 Scanning electron micrograph of the bacterial vegetation of a heart valve. (From McGowan, 1979, *Dental Update*, by permission)

If endocarditis is successfully treated, the healed valve is permanently scarred and thickened and such residual abnormalities make the patient highly vulnerable to episodes of reinfection.

A wide array of different bacterial species, fungi and chlamydiae can cause infective endocarditis. The most common causative organisms of endocarditis are shown in Table 12.2. More than 80% of infective endocarditis is caused by streptococci and staphylococci. The eminent position held by the *Strep. viridans* group of organisms in the league table (Table 12.2) indicates the major role played by the oral commensals in causing this life-threatening disease. *Strep. viridans* was responsible for about 90% of endocarditis cases some 50 years ago when most patients were young and rheumatic heart disease (related to *Strep. pyogenes* infection) was common. Although the patients seen today are older and the proportion of cases due to *Strep. viridans* (mainly *Strep. sanguis*) has fallen, this group of organisms is still the most frequent cause of endocarditis. It is noteworthy that nearly all patients with *Strep. viridans* endocarditis have a previous heart lesion and about a quarter give a history of a recent dental procedure as a precipitating factor.

Table 12.2 Causative microorganisms in infective endocarditis (cumulative data from several sources)

Microorganisms	Cases (%)
Total Streptococci	60
Streptococcus viridans	35
Streptococcus faecalis	13
Microaerophilic streptococci	3
Anaerobic streptococci	2
Others	7
Total Staphylococci	25
Staph. aureus	20
Staph. epidermidis	5
Miscellaneous	5
Culture negative	10

Factors which cause bacteraemia

As noted above, infective endocarditis is initiated when circulating microbes attach and proliferate within the fibrin thrombi attached to the damaged endocardium. As early as 1935 clinicians were aware that the oral cavity acts as a point of entry for organisms causing bacteraemia and that dental manipulations may set in motion the disease process. Since then many studies have shown that bacteraemia can occur following dental procedures such as extractions, surgical or non-surgical endodontics, gingivectomy, root-planing and scaling, and flossing. The frequency of bacteraemia has also been shown to be related to the preoperative oral sepsis of the patient and the degree of trauma and tissue injury.

172 Infections in medically compromised patients

With optimum blood culture technique a detectable bacteraemia can be shown to occur during tooth extraction in almost every case, tooth scaling in over a third of cases and endodontics with instrumentation beyond the apex in less than one third of cases. Other procedures, such as raising a mucoperiosteal flap, incision of abscesses and suture removal may cause bacteraemia but to a relatively minor degree.

The real risk of development of infective endocarditis in a 'risk' patient following dental procedures is difficult to ascertain. Whereas some suggest that less than 10% of infective endocarditis is attributable to bacteraemia of dental origin, others maintain this proportion to be as high as 90%. Furthermore, a proportion of infections are thought to be associated with random transient bacteraemias which commonly follow mastication, and even tooth brushing, in patients with chronic periodontitis.

Infective endocarditis prophylaxis

Since dental-related endocarditis may well be the most common potentially fatal complication of dental treatment, all practitioners must have a good working knowledge of the problem. There is good evidence that dentists know about endocarditis and its prevention. However, some of the prophylactic regimens which have been used in the past have been purely empirical and in recent years a more scientific approach has been used. Table 12.3 shows the key principles of prophylaxis.

Table 12.3 Key principles for antibiotic prophylaxis of patients at risk of developing infective endocarditis

Identification of 'at risk' patient
Patient awareness of risk status
Dental consultation in cardiac clinics
Preventive dental care
Antibiotic and antiseptic prophylaxis
Awareness of postoperative morbidity

Accurate identification of at risk patients
The main risk groups are patients with: (1) a history of congenital or acute rheumatic heart disease; (2) prosthetic heart valves; (3) a history of previous bouts of infective endocarditis; and (4) other rare cardiac conditions (see Table 12.1). Recently

it has been suggested that diabetics and alcoholics should also be regarded as risk patients. It is likely that this will unnecessarily complicate an area already difficult for dentists, since diabetes and alcoholism are likely to be cofactors of risk and not in themselves an indication for antibiotic prophylaxis. Because dentists usually identify a patient as at risk from his past medical history, it is important for them to obtain confirmatory and expert information from the patient's medical adviser. The doctor should also be able to advise if antibiotic prophylaxis is necessary or not. However, identifying risk patients by this means is totally dependent on the memory of the patient, which is not always reliable.

Patient awareness of risk status and dental involvement in cardiac clinics
There is evidence that patients do not always understand or remember a warning given to them about a rare complication of possible future dental treatment when they are involved in coping with a much more immediate life-threatening disease. Warning cards have been used with some success and a move towards the introduction of such a system nationally could give valuable information to dentists. It is important that dentists are part of the medical team involved in the pre- and postoperative management of patients undergoing cardiac surgery who are at risk of infective endocarditis.

Preventive dental care
The Working Party of the British Society for Antimicrobial Chemotherapy concluded that some but not all dental procedures could produce a bacteraemia capable of initiating infective endocarditis in a susceptible patient. The procedures in which a significant risk was thought to be involved were tooth extraction, surgery involving the gingival margin, and scaling. The risk associated with endocarditis developing from other dental procedures, as well as from regular oral hygiene or normal mastication, was considered to be very low and therefore acceptable. However, the need to administer prophylactic antibiotics to prevent endocarditis must be weighed carefully in each case and, if there is reasonable doubt, prophylaxis should normally be given. It follows that, ideally, susceptible patients should be exposed to risk procedures as rarely as possible and this can be best achieved by careful and intensive oral hygiene instruction, dietary advice, and regular dental examinations. The aim should be to reduce the amount of treatment for at risk patients to the

absolute minimum necessary for the maintenance of a healthy natural dentition for life. However, this ideal is not always realized and prophylaxis measures are required.

Antibiotic and antiseptic prophylaxis
Since the main source of microorganisms in significant dental bacteraemias is supra- and subgingival plaque, any local reduction in the numbers of such organisms before the start of treatment would be helpful. There is evidence that the incidence of detectable bacteraemias can be reduced by gently irrigating the gingival crevice area with 10 ml of either 1% chlorhexidine or 1% povidone–iodine, using a disposable plastic syringe fitted with a blunted needle. It must be emphasized that this topical antiseptic treatment should not be regarded as replacing antibiotic prophylaxis. However, since irrigation is safe, cheap, simple to use and reduces bacteraemia, it can be employed at all visits as an adjunct to antibiotic prophylaxis.

Awareness of postoperative morbidity
It is imperative that, even when antibiotic cover has been given, patients at risk should be instructed to report any unexplained illness. Infective endocarditis is often exceedingly insidious in origin and may develop months after the operation which might have precipitated it. Late diagnosis significantly increases the mortality and morbidity among the survivors.

Recommendations on antibiotic prophylaxis

A number of recommendations on antibiotic prophylaxis have been promulgated by the American Heart Association and the British Society for Antimicrobial Chemotherapy Working Party but the latter has gained most general acceptance by the dental profession in the UK (Figure 12.3).

The recommendation for the majority of risk patients, i.e. those who are not allergic to penicillin, who had not received penicillin more than once during the previous month, and required only local anaesthesia, is 3 g amoxycillin orally one hour before the operation, taken under supervision. A second 3 g dose 6–8 hours later has also been advocated as a result of experimental animal work.

Erythromycin is regarded as the best available alternative for oral use in patients who are allergic to penicillin, or where penicillin had been taken on a number of occasions within the

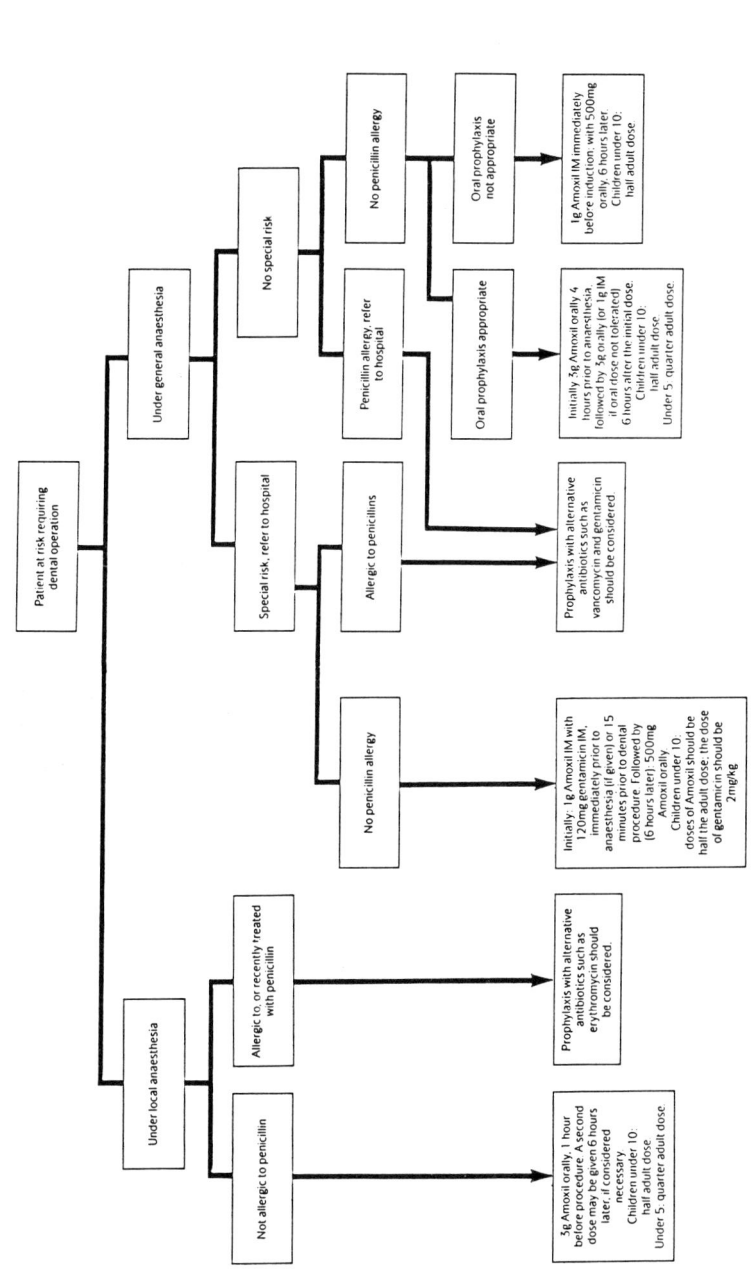

Figure 12.3 Endocarditis prophylaxis for dental procedures. (From *Amoxil in Dental Practice*, published by Bencard, by permission)

past month. For these patients erythromycin stearate 1.5 g orally should be given under supervision one to two hours before treatment, followed by a second dose of 0.5 g six hours later.

If a patient requires general anaesthesia, oral administration of drugs is contraindicated and prophylactic antibiotics need to be given by injection. The following regimen should be given to at risk patients with the exception of those with prosthetic heart valves or individuals who are hypersensitive to penicillin. The regimen consists of 1 g amoxycillin intramuscularly in 2.5 ml 1% lignocaine hydrochloride before induction and a further 0.5 g amoxycillin by mouth six hours later.

It is important for the practitioner to be aware that some categories of at risk patients need to receive prophylaxis in hospital and to be referred to the nearest consultant. The patients who should be referred are:

1. Patients who have received penicillin more than once in the previous month and require a general anaesthetic.
2. Patients who are allergic to penicillin and require a general anaesthetic.
3. Prosthetic valve patients requiring a general anaesthetic.
4. Patients who have had one or more attacks of infective endocarditis.

Details of prophylaxis for these patients are not discussed here since the recommendations involve the administration of gentamicin intramuscularly or gentamicin and vancomycin intravenously. The use of these regimens and drugs are outside the competence of a general dental practitioner.

In dentistry it is not uncommon for a second treatment visit to be necessary soon after the first. This raises the potential problem of the gingival crevice becoming colonized by significant numbers of bacteria resistant to amoxycillin, which could result in the subsequent failure of prophylaxis if the 3 g amoxycillin regimen was used. Recent studies have indicated that 3 g amoxycillin can be used safely on two occasions separated by a few days or weeks but that a recovery period of about one month should elapse before it is used again. If this delay is not possible, the erythromycin regimen should be used. In addition, the antibiotic sensitivity patterns of plaque flora in patients who have received recent antibiotic therapy can also be ascertained by *in vitro* laboratory tests. For this purpose, supra- and subgingival plaque samples related to the proposed surgical site(s) should be sent to the laboratory at least one week prior to surgery. The laboratory can then determine the overall sensitivity of plaque flora to a battery of antimicrobial agents from which the clinician could select an appropriate drug.

Prostheses

Heart valve prostheses

The dental management of patients with replacement heart valves can be undertaken by general practitioners, as described earlier for other at risk groups, as long as the patients require local anaesthesia, are not hypersensitive to penicillin and have not received penicillin within the past month. Otherwise the patient should be treated by the hospital service. Due to the very serious complications which can occur if prophylaxis fails in recipients of prosthetic valves, it may be felt that prophylaxis should be used during all other dental procedures in addition to those procedures described above.

Hip joint replacement

The situation with patients who have been fitted with other prostheses, e.g. hip joint replacement, is less certain. Since there is little evidence in the literature dealing with early or late postoperative infection of hip prostheses to suggest that bacteria derived from the mouth are involved, it is doubtful if prophylaxis is necessary. However, due to the serious and sometimes catastrophic effects of such infections in these patients, it has been suggested that all such individuals should receive antibiotic cover similar to that described earlier for patients at risk of developing endocarditis. It is important that the possible need and type of prophylactic cover should be discussed with the patient's doctor before dental treatment starts.

Immunocompromised patients

Mechanisms leading to immunocompromised states

Immunodeficiency disease can be either primary (developmental or genetically determined), which is rare, or secondary, due to procedures such as irradiation and cytotoxic drug therapy. Main causes of secondary immunodeficiency are shown in Table 12.4.

Rarely children are born with congenital deficiency of the immune system. These include deficiencies in B cells with depressed immunoglobulin production, T cell deficiency (e.g.

thymic aplasia), combined B and T cell deficiency, and neutrophil dysfunction.

A variety of diseases depress the immune system. In particular, neoplasms of the lymphoid system leading to lymphomas (Hodgkin's disease), leukaemia and multiple myeloma are by far the most important. Acquired immunodeficiency syndrome (AIDS) due to human immunodeficiency virus (HIV) is a good example of an infective disease which affects the immune system (see Chapter 15).

Table 12.4 The main causes of secondary immunodeficiency

Drugs:	Methotrexate
	Cytarabine
Malignant disease:	Acute leukaemia
	Hodgkin's disease
Infections:	AIDS
	Severe viral infections
Deficiency states:	Iron deficiency
Autoimmune disease:	Rheumatoid arthritis
Others:	Diabetes mellitus
	Down's Syndrome

A number of other diseases such as diabetes, renal failure, rheumatoid arthritis and autoimmune diseases (e.g. systemic lupus erythematosus), diminish immunity by often complex and incompletely understood mechanisms.

Modern medical treatment entails therapeutic modalities which either diminish or abolish immune function. Drugs which belong to this category and are extensively used include immunosuppressives, cytotoxics and steroids. Administration of these drugs, often in combination, to treat malignant diseases and to prevent graft rejection after organ transplantation predisposes to infection due to the dual effect of the drug as well as the disease itself.

Radiotherapy is widely used in cancer treatment and is a popular regime for therapy of head and neck cancer. In addition to the general depressive effect of radiotherapy on immune cells, it has localized effects on salivary glands and oral mucosa leading to mucositis and xerostomia of the oral cavity. The latter, as we shall see later, results in secondary oral infections.

Cofactors of importance

The types, presentation, severity and prognosis of infections in compromised patients are dependent on the interaction of a number of factors. These include: (1) the duration and depth of immunosuppression; (2) previous or current antimicrobial treatment (e.g. broad-spectrum antibiotics promote fungal infection); (3) the microorganisms to which the patient is exposed; (4) the degree of oral sepsis; and (5) the nature of the cytotoxic/immunosuppressive drug used (e.g. methotrexate in particular causes oral ulceration which may then become secondarily infected).

Oral and dental infections in compromised patients

Clinical presentation

The range of oral problems, which may be present in those who are immunodeficient due to radiotherapy, cytotoxic drugs, HIV infection and acute leukaemia, are shown in Table 12.5. As can be seen, some clinical conditions and infections are more commonly associated with a particular category of compromised patients than others.

Table 12.5 Microbiologically related oral problems in compromised patients

	Cytotoxic therapy	Radiotherapy	AIDS	Acute leukaemia
Mucositis	+	+	+	+
Ulceration	+	±	+	+
Xerostomia	±	+	±	−
Sialadenitis	−	+	±	−
Osteomyelitis	+	+	±	±
Candidosis	+	+	+	+
Herpes infection	+	−	+	±
Periodontal diseases	−	+	+	+
Dental caries	−	+	−	−

+, Usually present; ±, may be present; −, usually absent.

For example, in acute leukaemia the response to dental plaque is exaggerated with resultant gross gingival swelling. If untreated, the gingival margins become necrotic and ulcerated. However, periodontal disease is not a problem of such magnitude in patients on cytotoxic therapy as compared with other complications such as oral candidosis. However, the actual clinical presentation of these pathological changes in any given patient can vary within wide limits, depending on the various cofactors mentioned above, e.g. from mild mucositis alone to various mixtures of the nine categories shown in Table 12.5. The diagnosis of these conditions is not always possible on clinical grounds alone, partly because of uncertainty concerning the identity of the agents involved in conditions such as oral ulceration, salivary gland infections and osteomyelitis, and partly due to the altered way in which the defective defence mechanisms react with microorganisms, e.g. the absence of an acute inflammatory reaction. It is therefore particularly important to use laboratory tests during the diagnosis and management of these patients. The detailed management of oral infections seen in compromised patients is not mentioned in this chapter. However, the reader is referred to chapters on bacterial, viral and fungal infections (Chapters 8–10) and laboratory diagnosis (Chapter 13) for a fuller description of such diseases and their management.

Pretreatment management

Many of the oral problems shown in Table 12.5 can be drastically reduced in severity, if not prevented, by the use of careful pretreatment management.

Before radiotherapy or immunosuppressive drugs are used, a careful assessment of the patient's dental health should be performed and any restorative or surgical treatment carried out. Oral hygiene instruction and dietary advice is of paramount importance in reducing subsequent problems and the use of regular fluoride applications and chlorhexidine gel for dentulous, and 0.2% chlorhexidine mouthwash for edentulous patients should be used. Furthermore, the use of antifungal drugs prophylactically, such as nystatin pastilles just prior to and during treatment, can be advantageous. Necrosis of bone, particularly the mandible, due to irradiation endarteritis is a serious potential complication and, wherever possible, the extraction of non-restorable teeth and the removal of any other possible source of infection to irradiated bone should be carried out before treatment commences.

Management during treatment

However, it is not always possible to arrange adequate pretreatment management and in such a situation or where the patient first presents with oral problems, e.g. patients infected with

HIV, treatment may be arranged as shown in Table 12.6. It should be noted that, although there are a number of other methods and products which are available for the treatment of some of these oral and dental problems, it is outside the scope of this chapter to discuss these in detail. The interested reader is referred to the further reading section at the end of the chapter.

Table 12.6 An outline of dental management of oral problems in compromised patients

Oral complications	Principles of management	Chapter reference
Mucositis	Normal saline rinses; 0.2% aqueous chlorhexidine; 2% lignocaine viscous gel	—
Xerostomia	Saliva substitutes, e.g. carboxymethylcellulose	12
Sialadenitis	Laboratory investigation + antimicrobial agent	11
Osteomyelitis	Prevention by atraumatic extraction with antibiotic cover	6
Candidosis	Topical – nystatin (pastille/suspension) Systemic – flucytosine or others	9
Herpes infection	Severe – acyclovir (intravenous infusion) Other – acyclovir (tablets/cream)	10
Dental caries	Topical fluoride; dietary advice; 0.2% chlorhexidine mouthwash and gel	4
Periodontal disease	Oral hygiene; antiseptics and/or antibiotics	5

Due to the similarity of the clinical appearance of oral infective lesions in compromised patients, it is important that diagnosis and management are carried out with the assistance of laboratory tests, as described in Chapter 13.

The oral management of the patients must be closely linked with the medical treatment and it is essential that the dentist is regarded as part of the medical team and involved in any decisions related to oral problems. Usually the dentist responsible for patients in this situation will be a hospital dentist. However, in areas without dental schools and where the number of hospital dentists is small, some dental practitioners may be required to assist in the general dental management of compromised patients.

Xerostomia and Sjögren's syndrome

As has been mentioned above, one of the major oral problems seen in patients who receive cytotoxic and radiotherapy is dry mouth or xerostomia. Xerostomia may also be due to physiological and pharmacological effects, ageing and diseases of the salivary glands such as Sjögren's syndrome (primary and secondary). Consequent chronic dryness of the oral mucosa and the inadequate salivary cleansing mechanism increases the susceptibility of oral tissues to infection. Thus, increased incidence of caries, periodontal diseases, oral candidosis and ascending (bacterial) sialadenitis is seen in these patient groups.

Further, xerostomia may be associated with one or more of the following conditions: difficulty in eating and swallowing dry food; difficulty in wearing complete dentures; burning sensation of the oral mucosa; and changes in the sense of taste.

A drastic reduction or complete absence of salivary secretion in patients with Sjögren's syndrome has a profound effect on the normal oral flora. Much reduced moisture levels in the oral environment tend to favour growth of bacteria which are resistant to drying, such as *Staph. aureus*, and inhibit oral commensals adapted to high moisture levels. In addition, it is known that the pH of salivary secretions in these patients is low while the oxygen tension (E_h) is high, which may be unfavourable to the growth of bacteria such as *Veillonella*, commensal *Neisseria* and *Micrococcus* species. As a result, the number of these bacteria in Sjögren's patients is significantly reduced. Moreover, this environment favours the growth of *Candida* species, *Strep. mutans* and *Lactobacillus* spp. which are almost invariably isolated in large numbers from oral samples of Sjögren's patients.

Sequelae of chronic xerostomia

Extensive dental caries, which has been well documented in patients with Sjögren's syndrome, is a common sequel of chronic xerostomia. Dental caries is particularly seen in the cervical and incisal surfaces of teeth and frequently attacks the margins of dental restorations, even if located subgingivally. Thus prevention of dental caries is particularly important in all patients with xerostomia and this can be achieved by daily fluoride mouth rinsing, discontinuation of between-meals sucrose snacks, careful removal of dental plaque by proper, frequent brushing, and regular dental supervision. Severe caries may be controlled by application of fluorides. In addition, 2% chlorhexidine mouth washes will help to control gingivitis and other oral infections.

Candida species are considered as normal oral commensals of Sjögren's or xerostomic patients. Chronic atrophic candidosis, angular cheilitis, and papillary atrophy of the tongue are frequently seen in these patients as a possible consequence of candidal infection. The disease should be diagnosed and managed as discussed previously (Chapter 9). However, since the xerostomia is chronic, recurrence is common and each case must be followed closely for reappearance of the disease and treatment if necessary.

Acute complications such as ascending parotitis should be treated with antibiotics as described previously (Chapter 11). Pus should be sent for culture and antibiotic sensitivities and in the interim a penicillinase-resistant penicillin should be started.

Restoration of salivary function is an essential component of the management plan for patients with xerostomia. This could be done by stimulating salivary secretion with sialogogues or, if adequate amounts of saliva cannot be stimulated, using a saliva substitute (e.g. Saliva Orthana). Dental management of these problems is summarized in Table 12.6.

Further reading

Gould, J. M. (1988) Prevention of infective endocarditis. *British Dental Journal*, **164**, 31–32

Jaspers, M. T. and Little, J. W. (1984) Infective endocarditis: a review and update. *Oral Surgery, Oral Medicine and Oral Pathology*, **57**, 606–615

Little, W. J. and Falace, D. A. (1984) *Dental Management of the Medically Compromised Patient*, 2nd edn, Mosby, St Louis

MacFarlane, T. W., Ferguson, M. M. and Mulgrew, C. J. (1984) Post-extraction bacteraemia: role of antiseptics and antibiotics. *British Dental Journal*, **156**, 179–181

Scully, C. and Cawson, R. A. (1987) *Medical Problems in Dentistry*, 2nd edn, Wright, Bristol

Topazian, R. G. and Goldberg, M. H. (1981) *Management of Infections of the Oral and Maxillofacial Regions*, W B Saunders, Philadelphia

Part 4
Diagnostic oral microbiology

Chapter 13
Use of the microbiology laboratory

Clinical oral microbiology is a relatively new discipline and involves the study of specimens taken from patients suspected of having oral infections. The end result is a report which should assist the clinician in reaching a definitive diagnosis and help in deciding on therapy. It is clear from this that clinicians should be conversant with the appropriate techniques of specimen taking and comprehend at least the principles of microbiological analysis. The purpose of this chapter, therefore, is to outline the ways and means by which clinicians could make the optimal use of the laboratory. In addition, specific laboratory procedures such as sensitivity testing are described to assist clinicians to understand the laboratory reports which they receive.

Requesting a microbiology report

A spectrum of decisions and actions by a number of individuals are involved from the time the clinician decides to take a microbiological sample until he receives the laboratory report. This spectrum of events can be conveniently divided into three categories: (1) ordering, collecting and transportation of specimens; (2) microbiological laboratory processing and reporting; and (3) interpretation and use of information (Figure 13.1).

Clinical request

Factors directly influencing the quality of the sample include the clinical condition of the patient and current antibiotic therapy. If the patient is not suffering from a microbial infection then sampling for pathogens would be futile, for example, in patients with well-defined allergic reactions or traumatic lesions which are not secondarily infected. Successful isolation of pathogens from patients currently on antibiotic therapy would be difficult due to suppression of the pathogen(s) or due to qualitative and

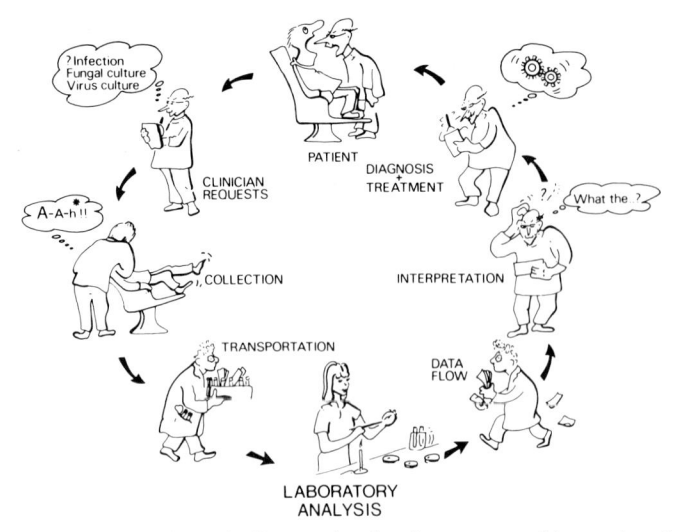

Figure 13.1 A schematic diagram showing the spectrum of interactions between clinicians and microbiologists

quantitative alterations in the commensal flora. Hence the importance of collecting microbial specimens prior to antibiotic therapy whenever possible. However, exceptions to this are where the patient is seriously ill, immunologically compromised or not responding to a specific antibiotic, in which case the necessity of obtaining an interim report as a guide to further management justifies such action.

Collection and transport of specimens

The resources of a microbiology laboratory can be efficiently utilized by sending appropriate specimens which are properly collected. Specimens should be as fresh as possible for optimal isolation of the pathogens. Many organisms, including some anaerobes and most viruses, do not survive for long in clinical specimens kept at room temperature. In addition, some organisms such as coliforms and staphylococci may multiply at room temperature and subsequent analysis of such specimens may lead to inaccurate and misleading results. Therefore, specimens should be transported as soon as possible to the laboratory in an appropriate transport medium. In the absence of a transport medium, dehydration of the specimen and exposure of the organisms to aerobic conditions occur, with the resultant death or reduction in their numbers. Various transport media are available and it is important to establish that the chosen medium is compatible with the microorganisms which are believed to be present in the clinical sample, e.g. a viral transport medium should not be used for transporting samples which are destined for bacterial culture, and prereduced transport media are required for strictly anaerobic bacteria. In circumstances when delay in transportation is inevitable, storage of the specimen at 4 °C for a short period may help to maintain the relative numbers and viability of organisms in the specimen.

When samples are transported by post they should be placed in robust, leak-proof containers and sent by first class letter post (not parcel post) with the legend 'Pathological specimens—handle with care' prominently displayed. Improperly packed specimens are an infective hazard for those who handle them. Information and containers for the postal transport of microbial samples can be obtained from the area medical/oral microbiology laboratory.

Provision of clinical information

The appropriate tests for each specimen received in the laboratory have to be selected by the microbiologist according to clinical information given on the accompanying request form. In brief, information such as age, main clinical condition, date of onset of illness, information about recent/current antibiotic therapy, antibiotic allergies, and history of previous specimens are all important to rationalize laboratory investigations and should be supplied with the specimen.

Investigations for some diseases require to be adequately discussed with the microbiologist beforehand in order to make the best use of laboratory resources. These include chronic infections where an actinomycotic or mycobacterial aetiology is suspected, tests for viral infections, and special investigations for specific antibiotics (e.g. assessment of minimum inhibitory concentration of antibiotics when treating infective endocarditis patients).

Laboratory analysis

In order to give the reader an insight into the laboratory analysis of specimens, a brief account of the analytical pathway of a pus specimen is given below and is shown in Figure 13.2.

As soon as a pus specimen is received in the laboratory it is given a laboratory identification number and inoculated on to blood agar for culture under aerobic and anaerobic conditions.

190 Use of the microbiology laboratory

Two further plates (one aerobic and the other anaerobic) are set up to assess the primary antibiotic sensitivities of the putative pathogens. Finally a smear of pus is made on a glass slide, heat dried and Gram stained for microscopic examination.

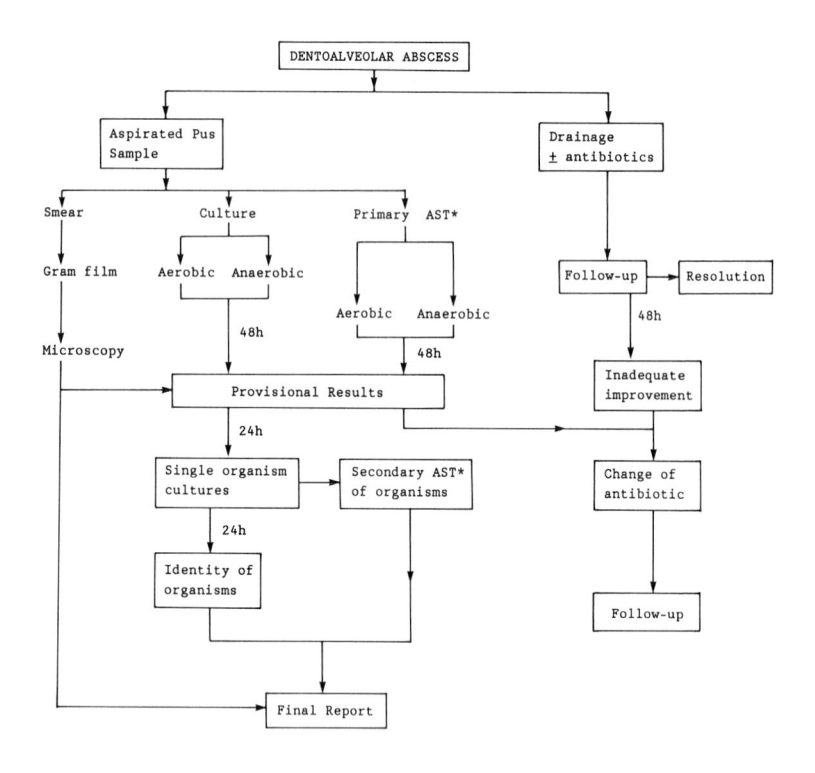

Figure 13.2 Analytical flow-chart of a pus specimen showing interaction between the laboratory and the clinician. * Antibiotic sensitivity test

As most of the oral pathogens are slow-growing anaerobes it is generally accepted that an incubation period of 2 to 3 days at 37° is required to produce visible colonies of these organisms. After 48 hours the plates are inspected for growth and if a single colony type is present its morphology is determined and identified. However, as a majority of dentoalveolar infections are polymicrobial in nature, samples usually yield a mixture of organisms which require to be isolated in pure culture to enable identification. The primary sensitivity plates are examined at

48 h or earlier (24 h) if the clinician requests an urgent sensitivity report.

At this stage, when about 2–3 days have elapsed, the microbiologist will be in a position to give the clinician an overall primary sensitivity result and presumptive identification of the organisms present. It may take a further 2–3 days to issue the final report as this is the period required for identifying bacteria and for computing reports.

The final report and its interpretation

Generally speaking, the clinician should receive the final report within a few days after the delivery of the specimen to the laboratory. However, in certain situations, e.g., when virus culture is requested, this may take a longer period. In this situation the clinician should make a telephone enquiry for a provisional verbal report. This is of particular help when the patient's condition is not improving and the clinician can discuss the problem with the clinical microbiologist with regard to a possible change in the antimicrobial therapy and management, or the need for additional specimens. Sometimes the final report may never reach the clinician due to an inefficient data flow system or when more than one clinician is involved in the management of the patient.

The interpretation of the final report by the clinician may sometimes be difficult, especially in answering the question: 'What does it mean?'. There is much confusion when, as is often encountered in oral specimens, the organism reported as the putative pathogen is a member of the commensal oral flora. Such problems of interpretation mainly arise in circumstances where the quality of the specimen, the history of its transport, and clinical data are unsatisfactory. Whenever there is a particular difficulty with such interpretation or, indeed, when choosing an appropriate antibiotic, good collaboration between the clinician and the clinical microbiologist is of the greatest importance.

Specimen collection from oral infections

Bacteriological specimens

Dentoalveolar abscess
It has been the practice for many years to sample the pus from a dentoalveolar abscess by incising the abscess (usually intraorally) and collecting the pus on a swab. Such a specimen is usually

grossly contaminated with resident oral commensal microorganisms. Although the manner in which a specimen is taken is dependent to some extent on the nature of the abscess, the optimal way to collect pus is by needle aspiration. The mucosal surface is first isolated and decontaminated with 0.2% chlorhexidine gluconate (Hibitane); patient discomfort at this stage or during subsequent needle aspiration can be relieved by surface anaesthesia using an ethyl chloride spray. Ideally, an aspirated pus specimen should be transported rapidly to the laboratory in the recapped syringe itself. Contamination of sample and exposure of organisms to aerobic and dehydrating conditions is minimized by this procedure although care should be taken when recapping to avoid a needle-stick injury.

It is important to note that aspiration of pus *per se* is not a satisfactory method of achieving drainage: subsequent incision of the abscess according to surgical principles may be required in some situations in order to drain the residual pus.

When aspiration fails to yield any material, incision and collection of pus is the only available option. In order to minimize contamination the area should be isolated and disinfected with 0.2% chlorhexidine before collection. In spite of these preventive measures, such samples almost invariably become contaminated.

Collection of parotid pus
In general, collection of uncontaminated parotid pus for microbiological examination is difficult due to: (1) relatively poor accessibility of the parotid duct; and (2) the absence of copious amounts of pus in the lesion. Therefore, two main techniques are used in collection of parotid pus.

In the presence of adequate quantities of pus it can be collected by aspiration through a small bore plastic catheter (attached, aseptically, to a sterile 2 ml syringe) inserted into the duct. The pus should then be transferred to the clinic in the syringe itself. The second method entails injecting about 0.5 ml sterile saline via the catheter and the 'saline wash' of the parotid duct system is then aspirated and sent to the laboratory as described above. This method is particularly useful when there is no frank discharge of pus, although care should be taken not to instil a large volume of saline and to carry out this procedure with minimal pressure to prevent further spread of infection within the gland.

In situations where the above procedures cannot be carried out then pus can be collected on to a swab by milking the duct.

However, to prevent contamination the duct orifice and the surrounding mucosa should be wiped with 0.2% chlorhexidine, then dried, prior to swabbing.

Material from a fistula should also be collected by using one of the above techniques appropriate for the particular clinical situation.

Specimens from mucosal and skin lesions
It is almost impossible to collect specimens from superficial intraoral sites without contamination from the indigenous salivary flora. Some authorities, however, recommend repeated washing of the lesion with sterile saline or water, isolating the area with cotton wool swabs, drying the area and collecting the specimen by either firm swabbing or scraping the lesion with a flat plastic instrument.

Specimens from perioral skin and angular cheilitis lesions should be sampled with a swab moistened in sterile saline. The swab should be gently applied to the lesion and rolled over to gather any putative pathogens. Particular care should be taken to minimize contact of the swab with the surrounding commensal skin flora.

Specimens from periodontal infections
In general it is very difficult to collect uncontaminated specimens from patients with gingival and periodontal lesions. This may explain to some extent the reason why the microbial aetiology of chronic periodontal diseases remains unclear. However, various methods have been used to sample subgingival plaque with some degree of success (e.g. paper point). None of these techniques are ideal and further information is given in Chapter 5 and in the references at the end of this chapter.

However, there are two situations where the microbiological analysis of a sample from a periodontal lesion can be of particular value to the clinician. Firstly, when a clinical diagnosis of acute ulcerative gingivitis (AUG) is uncertain in an atypical case, microscopic examination of material from the lesion can confirm or refute the clinical diagnosis. For this, a deep gingival plaque sample should be obtained and a smear prepared. The smear, after heat fixation, is Gram stained and observed under the microscope. A positive smear will reveal numerous spirochaetes, fusobacteria and polymorphonuclear leucocytes, confirming a clinical diagnosis of AUG (see Chapter 5). Although a negative smear may show small numbers of some of these elements, Gram-positive and negative cocci and bacilli predominate. It is imperative that the superficial plaque is removed before the sample is taken, otherwise a false negative result may occur. The main advantages of examining smears from AUG lesions are rapid confirmation of clinical diagnosis and usefulness in monitoring the resolution of the disease.

The second situation where a microbiological investigation may be of value is the diagnosis and management of localized juvenile periodontitis, rapidly progressive periodontitis and refractory adult periodontitis.

Tissues sent for microbiological examination
Tissues which are removed at operation or post-mortem may sometimes require to be sent to the laboratory for microbiological investigation. If the tissue is sent immediately to the laboratory then transport medium is not required. A sterile Petri dish with a sealed lid can be used for transportation and this minimizes dehydration and contamination. It is important that, whenever possible, the clinician informs the laboratory in advance of sending tissue specimens since they are processed in a different way from routine specimens.

Endodontic specimens (see Chapter 7)

Virology specimens

Collection of specimens for virology
Specimens for viral culture should be collected as soon as possible following the onset of disease. The lesion is sampled by rubbing a sterile cotton wool swab over the affected area and then immediately placing the swab into viral transport medium (VTM). As the bottle containing VTM cannot accommodate the whole swab it should be broken level with the top of the bottle and the screw capped firmly to prevent any leakage of the medium in transit.

Viral transport medium is a general term to describe a solution containing proteins and balanced salts which stabilizes the virus during transportation. Antibimicrobial agents are also added to kill any bacteria present in the sample. If there is a delay in transportation of the swab in VTM it should be stored at 4 °C for up to 18 hours.

Laboratory isolation and identification of viruses
A virological investigation may be useful in dentistry for several

purposes, for instance: (1) as an aid to the diagnosis of specific lesions in an acutely ill patient; or (2) for the serological evaluation of the hepatitis B status of a patient which has public health implications in terms of dental care delivery.

Laboratory procedures for the diagnosis of viral infections are of three main types: (1) direct microscopic examination of host tissues for characteristic cytopathological changes or for the presence of viral antigens; (2) isolation and identification of virus from tissues, secretions or exudates; and (3) detection of virus specific antibodies or antigens in serum samples.

Direct microscopy of clinical material
Members of several virus groups including the herpes group induce characteristic histopathological changes that can be the basis of a presumptive diagnosis. Herpes viruses produce group specific individual intranuclear inclusion bodies but viruses must be identified using additional tests. Infection with either herpes simplex virus or varicella zoster virus results in the production of intranuclear inclusion bodies and multinucleated giant cells as a result of cell fusion. These two features can be seen in a stained smear of fluid or scrapings from a freshly ruptured vesicle. In recent years there has been considerable progress in the development of microscopical techniques for the detection of virus and viral antigens in tissues using immunological techniques. These include the use of fluorescent antibody, and immunoperoxidase techniques. Electron microscopy of negatively stained smears also is used in viral diagnosis.

Isolation and identification of viruses
For the isolation and subsequent identification of a virus it must be cultivated in living cells. Since no single type of host cell will support the growth of all viruses, a number of different methods of culturing viruses have been developed: (1) tissue culture cells (e.g. monkey kidney cells, baby hamster kidney cells); (2) embryonated eggs; and (3) laboratory animals such as suckling mice. Tissue culture cells, however, are the cheapest and most widely used system for isolating viruses.

After inoculation of a monolayer of tissue cells with a clinical sample diluted in VTM, it is examined daily for microscopic evidence of viral growth and assessed each day for about 10 days. Viruses produce different kinds of degenerative changes, termed cytopathic effect (CPE), in susceptible cells. The cell type supporting virus growth and the nature of CPE produced help virologists to identify individual viruses. For instance herpes viruses growing in monkey kidney cells produce fused cells in which nuclei aggregate to form multinucleate giant cells. Finally, the integrity of the monolayer is disrupted resulting in the appearance of foci of rounded cells. The type of CPE and other data related to cell culture of viruses affecting the orofacial region is shown in Table 13.1. The rate at which the CPE is seen can vary, depending on the virus strain and the concentration of the inoculum, from 24 hours up to several days.

Table 13.1 Susceptibility of cell cultures for viruses which infect oral and perioral regions

Virus	PMK	HFF	HFL	Nature of CPE	Time of appearance
Herpes simplex	+	+	+	Rounding	1–5 days
Varicella zoster	−	+	+	Focal lesions	1–4 weeks
Cytomegalovirus	−	+	+	Focal lesions	1–4 weeks
Mumps	+	−	−	Syncytia/none	7–10 days
Coxsackie A[a]	−	−	±	Rounding, sloughing	1–2 weeks
Coxsackie B	+	−	+	Rounding, sloughing	4–9 days

PMK, primary monkey kidney; HFF, human foreskin fibroblasts; HFL, human fetal diploid lung; CPE, cytopathic effect.
+, Susceptible; −, not susceptible; ±, slightly susceptible.
[a] Some strains grow only in suckling mice.

Methods for the subsequent identification of virus isolates include: (1) electron microscopical examination; (2) haemadsorption and haemagglutination inhibition tests; (3) growth neutralization assays using virus specific antiserum; and (4) specific fluorescent antibody techniques.

Serodiagnosis of viral infections
Many virus infections, as described in Chapter 10, have a short prodromal illness during which virus shedding from the oropharynx is maximal. After the onset of clinical symptoms, the rate of virus shedding falls rapidly and by about 5 days from the onset of symptoms, virus shedding has ceased. It is, therefore, rarely worth sending material for virus isolation after this period.

Under these circumstances serodiagnosis of viral infections is the other available option. Successful serodiagnosis of viral infections requires the timely collection of a pair of blood samples, one in the acute and the other in the convalescent phase of the disease. Obviously, acute phase serum should be collected as early as possible when a viral illness is suspected while convalescent phase serum is collected when the patient has recovered, usually some 10–20 days after the first specimen. If the interval between collecting the two samples is reduced, a lower rate of positive findings may result.

Serological test results are interpreted by comparing the antibody titres of the acute and convalescent sera. A titre is a term used to describe the reciprocal of the highest serum dilution which shows antibody activity in the test. A greater than four-fold rise in titre between the acute and convalescent samples is generally considered to be a positive result, indicating the patient has had an acute illness due to the specific virus.

Complement fixation tests are most commonly used in serodiagnosis. However, increasingly often, the demonstration of IgM antibody specific for a particular antigen in serum samples is being used (Figure 13.3). The advantages of this method over

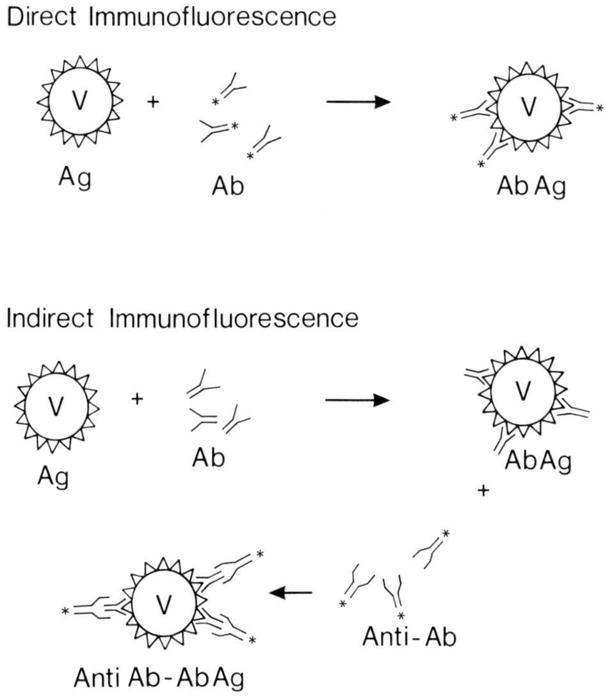

Figure 13.3 Principles of direct (one-step) and indirect (two-step) immunofluorescence techniques used in detection of viral antigens. * Immunofluorescence label. V, viral antigen; Ag, antigen; Ab, antibody

the comparison of viral titre in paired sera described above are that only a single serum sample is needed and results are available in a relatively short time. Virus specific IgM can be identified for herpes simplex, herpes zoster, Epstein–Barr, cytomegalovirus and mumps infections (see Chapter 10). Immunological and tissue culture detection methods have been successfully combined in order to shorten the time required to identify viral infections.

Other methods frequently used in serodiagnosis of viral infections include haemagglutination, radioimmunoassay and enzyme-linked immunosorbent assay (ELISA) tests.

Serodiagnosis using multiple antigen systems
Some viruses, such as mumps virus and hepatitis B virus, present with more than a single antigen which appear at different periods of the illness. This feature can be exploited to detect the state of illness by using a single sample of serum without waiting for convalescence. A variety of antigens and antibodies used in the detection of various phases of hepatitis B virus infection is described in some detail in Chapter 15.

Diagnosis of fungal infections

The most frequent fungal infection the dental practitioner has to diagnose is oral candidosis. As discussed in Chapter 9, oral candidosis manifests in a variety of ways and it is important to distinguish this condition from other diseases, such as leukoplakia and lichen planus, which may resemble them. Mycological investigation for candidosis is also important to monitor therapy and to establish the presence of persistent reservoirs of infection (e.g. prostheses).

A number of sampling methods have been used to investigate oral candidosis. However, sampling the lesion with a dry swab is still the conventional method of diagnosis. Smears taken immediately after swabbing should accompany the swab (see Table 9.3). A smear is taken by scraping the lesion with the edge of a flat plastic instrument and transferring the sample to a glass microscope slide. In the laboratory, the smear is stained with Gram's stain or periodic acid–Schiff (PAS) reagent and examined microscopically to visualize the hyphae or blastospores of *Candida*. Their presence in large numbers suggests active infection (Figure 13.4).

Once the swabs are received in the laboratory they are

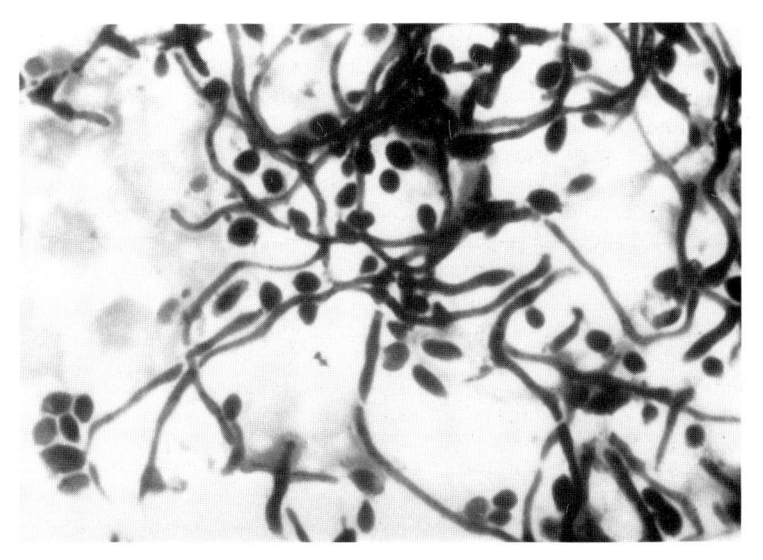

Figure 13.4 A Gram-stained smear of a palatal lesion showing blastospore phase (yeast phase) and hyphal phase of *Candida*. Magnification ×1000, reduced to 60% on reproduction

cultured on Sabouraud's medium and incubated for 48–72 hours, when *Candida albicans* appear as cream coloured large convex colonies. Yeasts are speciated by sugar fermentation and assimilation tests and the 'germ-tube' test. The latter is a useful quick test to differentiate *C. albicans* from the other *Candida* species such as *C. glabrata* and *C. krusei*. When a small inoculum of the isolated yeast is incubated in serum at 37 °C for about 3 hours, almost all strains of *C. albicans* produce short, cylindrical extensions termed 'germ tubes' as opposed to the other *Candida* species which do not exhibit this characteristic.

Swabs and smears, although useful for assessment of the presence or absence of *Candida* in specific areas of the mouth, cannot provide a quantitative estimate of oral yeast carriage. However, oral rinse culture and imprint culture techniques permit quantitation of yeast populations and are useful in screening patients for oral yeast carriage. They are less useful for differentiating commensal yeast carriage from clinical candidal infection and, if infection is suspected, swabs and smears from lesions should be obtained.

An oral rinse sample is obtained by requesting the patient to rinse his mouth thoroughly with 10 ml of sterile physiological saline for 1 minute. The sample is then expectorated back into a universal container and transported to the laboratory where quantitative estimations of yeast numbers can be carried out.

In imprint sampling a sterile square foam pad (2 × 2 cm) is used to obtain a sample by pressing the pad firmly on the infected or normal mucosal surface. The pad is then placed on Sabouraud's agar medium and incubated to assess the number of yeast colonies present. This method is particularly useful in assessing the yeast populations inhabiting various niches of the oral cavity, such as the buccal mucosa, tongue and the palate.

Incisional and excisional biopsies are useful clinical techniques in the diagnosis of persistent oral white lesions which are thought to be related to infection with *Candida* species. As a significant proportion of chronic candidal leukoplakic lesions are premalignant, a biopsy in addition to a swab is essential (Chapter 9).

Serological diagnosis

A number of new serological tests are available to diagnose systemic candida infections but these are of little or no value in laboratory diagnosis of superficial infections of the mucous membranes.

Other laboratory investigations

On occasions chronic candida infections are associated with nutritional and haematological abnormalities and appropriate laboratory investigations should also be carried out.

The use of laboratory investigations in the management of antimicrobial therapy

The clinical microbiologist can on occasions give a presumptive diagnosis of a disease and suggest suitable treatment by examining a Gram film of material prepared from the lesion. In other circumstances, once certain organisms have been isolated and identified, antimicrobial sensitivity can sometimes be predicted, e.g. strict anaerobes are usually sensitive to metronidazole and *Candida albicans* is almost always sensitive to both nystatin and amphotericin B. However, it is essential to base rational therapy on the results of laboratory antibiotic tests.

Susceptibility of organisms to antimicrobial agents

In clinical microbiology a microbe is susceptible to an antimicrobial agent if it is inhibited by a concentration of the drug normally obtained in human tissues after a standard therapeutic dose. The reverse is true for a resistant organism. Organisms are considered intermediate in susceptibility if the inhibiting concentration of the antimicrobial agent is slightly higher than that obtained with a therapeutic dose.

Laboratory testing for antimicrobial sensitivity

The action of an antimicrobial drug against an organism can be measured qualitatively (disc diffusion tests), semi-quantitatively (break-point tests), or quantitatively (MIC or MBC tests, see below). These *in vitro* tests indicate whether the expected therapeutic concentration of the drug given in standard dosage inhibits the growth of a given organism *in vivo*.

Laboratory results can only give an indication of the activity of the drug *in vitro*, and its effect *in vivo* depends on factors such as the ability of the drug to reach the site of infection and the immune status of the host. A strong host defence response may give the impression of 'successful' drug therapy, even although the infecting organism was 'resistant' to a specific drug when laboratory tests were used.

Disc diffusion tests

The most commonly used method of testing the sensitivity of a microorganism to an antimicrobial agent is the disc diffusion test. In this technique, both test and control organisms are inoculated on to the same sensitivity agar plate, as shown in Figure 13.5. The 'Oxford staphylococcus' is used as a control in most tests as it is sensitive to most antimicrobial agents at therapeutic drug concentrations. The drug-impregnated filter paper discs are placed between the control and test zone. After 18 hours incubation at 37 °C, a zone of growth inhibition is observed surrounding the disc, depending on the sensitivity of a particular organism to a given agent. The zones of growth inhibition can now be compared with the controls and the organism reported as sensitive, resistant, or moderately sensitive to the antimicrobial agent, as shown in Figure 13.5.

Antimicrobial sensitivity tests of this type can be divided into primary (direct) and secondary (indirect). A primary test is carried out by inoculating the clinical sample, say pus, directly

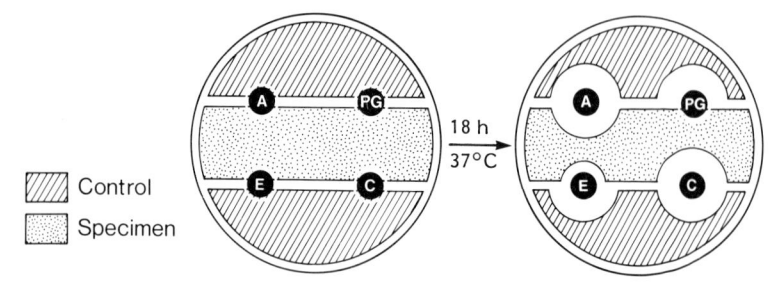

Figure 13.5 Disc diffusion (Stoke's) method of antibiotic sensitivity testing. Sensitive control (Oxford staphylococcus) is spread over top and bottom thirds of plate. The specimen (or test organism) is spread over middle third. This particular test organism is sensitive to ampicillin (A) and cephaloridine (C), moderately sensitive to erythromycin (E) and resistant to penicillin (PG)

on to the test zone of the plate. The advantage of this is that the overall sensitivity results for the organisms present in pus will be available after 24–48 hours incubation. This is particularly useful when treating debilitated patients with acute infections, e.g. dentoalveolar abscesses. Secondary sensitivity tests are performed on a pure culture of the isolates originally present in the specimen but the results are not available for at least 2–4 days after sampling (see Figure 13.2).

Assessment of minimum inhibitory concentration (MIC) and minimum bactericidal concentration (MBC)
These tests give a quantitative assessment of the potency of an antibiotic. A range of two-fold dilutions of an antimicrobial agent can be incorporated into a suitable broth in a series of tubes (tube dilution technique). The broth is inoculated with a standardized suspension of the test organism and incubated for 18 hours. The minimum concentration of the drug which inhibits the growth of the test organism in the tube is recorded as the MIC. Subsequently, a standard inoculum from each of the tubes in which no growth occurred may be subcultured on to blood agar to determine the minimum concentration of the drug required to kill the organism (MBC). The MBC is defined as the minimum concentration of drug which kills 99.9% of the test microorganisms in the original inoculum. A variation of the MIC test, called the 'breakpoint' or 'critical' concentration test is a semi-quantitative version of the standard MIC test. In this method, which may be carried out by incorporating antimicrobial agents into either broth or agar, a limited number of drug concentrations are used rather than an entire series of doubling

dilutions. These tests are not routinely performed but are useful in patients with serious infections where optimal antimicrobial therapy is essential. Examples include streptococci isolated from blood cultures from patients with infective endocarditis and some bacterial strains causing septicaemia in immunosuppressed patients.

Further reading

Howie, J. (1979) Clinicians and microbiologists should get together. *Journal of Infection*, **1**, 19–22

Mintz, G. A. and Rose, S. L. (1984) Diagnosis of oral Herpes simplex virus infections: practical aspects of viral culture. *Oral Surgery, Oral Medicine and Oral Pathology*, **58**, 486–492

Samaranayake, L. P. (1987) The wastage of microbial samples in clinical practice. *Dental Update*, **14**, 53–61

Stokes, E. J. and Ridgway, G. L. (1988) *Clinical Bacteriology*, Edward Arnold, London

Chapter 14
Antimicrobial chemotherapy

Antimicrobial agents include antibacterial, antiviral, antifungal and antiprotozoal agents. Apart from the last group, dental practitioners are frequently called upon to prescribe these drugs and therefore should be conversant with the mode of action, spectrum of activity, side-effects and contraindications of at least the major groups of antimicrobials. This chapter summarizes the principles of antimicrobial therapy together with notes about the nature and use of antimicrobials frequently prescribed in dentistry.

All antimicrobials demonstrate selective toxicity, in that the drug can be administered to humans with reasonable safety while having a marked lethal or toxic effect on specific microbes. The corollary of this is that to some extent all antimicrobials have adverse effects on humans and should therefore be used rationally and only when required.

Bacteriostatic and bactericidal antimicrobial agents

Antimicrobial agents are classically divisible into two major groups as bactericidal agents which kill bacteria or bacteriostatic agents which inhibit multiplication without actually killing the pathogen. However, the distinction is rather hazy and is dependent on factors such as the concentration of the drug and the organism concerned and severity of infection. For instance, it is known that erythromycin is bacteriostatic at low concentrations and bactericidal at high concentrations.

Further, it is important to realize that host defence mechanisms play a major role in eradication of microbial pathogens from the body and it is not essential to use bactericidal drugs to treat most infections. A bacteriostatic drug which arrests the multiplication of pathogens and so tips the balance in favour of the host defence mechanisms is satisfactory in many situations.

Mode of action of antimicrobial agents

Antimicrobial agents inhibit the growth of or kill microorganisms by a variety of mechanisms. In general, however, one or more of the following four specific target sites are involved: (1) the cell wall, (2) the cytoplasmic membrane, (3) ribosomes, and (4) nucleic acid replication sites. The detailed modes of action of antimicrobials are outside the scope of this book and the reader is referred to texts in the reading list for this information. A summary of the mode of action of commonly used antimicrobials in dentistry is given in Table 14.1

Table 14.1 Mode of action of antimicrobials commonly used in denistry

Target site	Drug	Cidal/Static	Comments
Cell wall	β-Lactams, e.g. penicillin, ampicillin, cephalosporin, cloxacillin	Cidal	Interfere with cross-linking of cell wall peptidoglycan molecules
	Bacitracin (topical)	Cidal	Inhibit peptidoglycan formation
Ribosomes	Erythromycin Fusidic acid (topical)	Static[a] or Cidal[b]	Interfere with translocation, thus inhibiting protein synthesis
	Tetracycline	Static	Interferes with attachment of transfer RNA thus inhibiting protein synthesis
Cytomplasmic membrane	Polyenes, e.g. nystatin, amphotericin	Static	Disrupts yeast cell membrane
Nucleic acid replication	Metronidazole	Cidal	Interferes with DNA replication
	Idoxuridine Acyclovir	Cidal	Interferes with DNA synthesis in DNA viruses

[a] Low concentrations; [b] high concentrations.

General principles of antimicrobial therapy

General indications

Antimicrobial agents should be prescribed on a rational clinical and a microbiological basis. In general, therapy should be considered for patients with orofacial infections when one or more of the following conditions are present: (1) patients with fever who may complain of malaise, fatigue, dizziness and weakness; (2) spreading infection without localization; (3) chronic infection despite drainage or débridement; (4) medically compromised patients; and (5) cases of osteomyelitis, bacterial sialadenitis and acute ulcerative gingivitis.

The choice of drug

The choice of drug is strictly dependent upon the nature of the infecting organism(s) and their sensitivity pattern(s). However, in a clinical emergency such as septicaemia or Ludwig's angina, antimicrobial agents must be prescribed empirically until laboratory tests have been completed. In other conditions such as acute ulcerative gingivitis where a Gram-stained smear of the lesion is diagnostic, therapy with metronidazole should be instituted without further laboratory tests since it is extremely difficult to isolate and identify spirochaetes (see Chapter 5). In general another antimicrobial should be prescribed if the patient has had penicillin within the previous month. This is because of the natural selection and persistence of resistant bacterial populations when they are exposed to the drug.

Spectrum of activity of antimicrobial agents

Antimicrobial agents can be categorized into broad-spectrum and narrow-spectrum drugs depending on their activity against a range of Gram-positive and Gram-negative bacteria. Penicillin is a good example of a narrow-spectrum antibiotic with activity mainly against the Gram-positive bacteria. Similarly, metronidazole is a recent example of a narrow-spectrum drug which acts almost entirely against strict anaerobes and some protozoa.

On the other hand, broad-spectrum antimicrobial agents are active against many Gram-positive and Gram-negative bacteria; examples include tetracyclines, ampicillins and cephalosporins. Hence the broad-spectrum agents are often used for the blind or empirical treatment of infections when the likely causative pathogen is unknown. This, unfortunately, leads to abuse of broad-spectrum agents with consequential emergence of resistant organisms which were originally sensitive to the drug. The spectra of activity of some broad- and narrow-spectrum antimicrobial agents used in dentistry are shown in Table 14.2.

Table 14.2 Spectra of antimicrobials commonly used in denistry

Drug	Spectrum
Penicillin V	1. Aerobic Gram-positives (e.g. streptococci, pneumococci, β-lactamase-negative staphylococci, lactobacilli, actinomyces) 2. Anaerobic Gram-positives (e.g. anaerobic streptococci) 3. Anaerobic Gram-negatives (e.g. most *Bacteroides*, fusobacteria, *Veillonella*)
Penicillinase-resistant penicillins (e.g. flucloxacillin)	All the above, including β-lactamase-producing staphylococci
Ampicillin	As for penicillin, also includes *Haemophilus* spp.
Cephalosporins	As for penicillin, also includes some coliforms
Erythromycin	Gram-positives mainly but some anaerobes not susceptible at levels obtained by oral administration
Tetracycline	Broad spectrum. Many Gram-positives and negatives
Metronidazole	All strict anaerobes are sensitive, including some protozoa. Of questionable value for facultative anaerobes

Combination therapy

Whenever possible, a single antimicrobial agent should be used to reduce the incidence of possible side-effects, the emergence of resistant bacteria, and drug costs. However, there are certain clinical situations where a combination of drugs is valuable, e.g. the necessity of achieving a high bactericidal level when treating patients with infective endocarditis. Other examples include the use of gentamicin and metronidazole in the empirical treatment of a patient with serious abdominal sepsis, and combined drug therapy in the management of tuberculosis.

Antimicrobial prophylaxis

Antimicrobial prophylaxis has been defined as the use of a drug to prevent colonization or multiplication of microorganisms in a susceptible host. The value of prophylaxis depends upon a balance between the benefit of reducing the infection risk and consequent secondary morbidity against the possible toxic effects to the host, alteration of the host commensal flora, the emergence of resistant bacteria and cost. When used rationally, prophylaxis can reduce morbidity and the cost of medical care although when used inappropriately a false sense of security with increased costs is likely to result.

The functions of antimicrobial prophylaxis are: (1) to prevent the colonization of the host by a small number of virulent agents; or (2) to prevent implanted organisms reaching a critical mass sufficient to produce infection; or (3) to prevent the emergence of latent infection.

In dentistry, antibiotics are used as prophylactic agents in a number of situations, although in some instances their use is highly controversial as the efficacy of prophylaxis has not been proven scientifically.

Prophylactic antimicrobial agents are definitely indicated during dental treatment involving patients with at risk status with respect to infective endocarditis (see Chapter 12), facial fractures or compound skull fractures, and cerebral rhinorrhoea. Oral surgery performed in immunocompromised patients, or in patients who have recently received radiotherapy to the jaws, may also require prophylaxis. In the former situation the patients are susceptible to infection due to generalized immunodeficiency while in the latter instance the jaws succumb to infection due to severe ischaemia of the bone consequential to radiotherapy.

Prophylactic antimicrobial agents may possibly be required in patients with prosthetic hip replacements and ventriculoatrial shunts. Consideration should also be given to prophylactic drugs during insertion of implants or bone grafting. However, the advantages and disadvantages of antimicrobial therapy in such instances should be carefully weighed in each individual case. The routine administration of antimicrobial agents after uncomplicated third molar surgery is often given but is probably unnecessary.

Prescribing an antimicrobial agent

The following questions should be considered before any antimicrobial agent is prescribed.

Is there an infective aetiology?

When there is no good clinical evidence of infection antimicrobial agents are rarely necessary, except in prophylaxis which is discussed above.

Have relevant specimens been taken before treatment?

As explained in Chapter 13, appropriate specimens should be collected before drug therapy is started as the population of pathogens may be reduced, and therefore less easily isolated, if specimens are collected after antimicrobial agents have been taken. Further, the earlier the specimens are taken, the more chance that the results will be available to help with patient management.

Timing the start of treatment

In patients with life threatening infections e.g. Ludwig's angina, often intravenous therapy should be instituted immediately after specimen collection. Antimicrobial therapy may be withheld in chronic infections such as angular cheilitis until laboratory results are available.

Which antimicrobial agent?

It is important to consider, as described below, the pharmacodynamic effects, including toxicity, when choosing a drug from a number of similar antimicrobial agents which are available to treat many infections. Furthermore, an adequate medical history, especially in relation to past allergies and toxic effects, should be taken prior to therapy.

Pharmacodynamics of antimicrobial agents

Dosage

Antimicrobial agents should be given in therapeutic doses sufficient to produce a tissue concentration greater than that which is required to kill or inhibit the growth of the causative microorganism(s).

Duration of treatment

Ideally, this should be long enough to eliminate all or nearly all of the pathogens as the remainder will in most instances be destroyed by the host defences. In practice this can not be precisely timed and standard regimes last for some 3 – 5 days, depending on the drug. However, recent studies have shown that a short-course, high-dose therapy of certain antibiotics (e.g.

amoxycillin) is as effective as a conventional 5-day course. The other advantages of short courses of antimicrobial agents are good patient compliance and the minimal disturbance to commensal flora with an associated reduction in side-effects such as diarrhoea.

Route of administration

In seriously ill patients the route of choice is parenteral to overcome problems of absorption from the intestinal tract. All antimicrobial agents given by mouth must be acid stable.

Distribution

The drug must reach adequate concentrations in the focus of infection. Some antibiotics, such as clindamycin, which penetrate well into bone are preferred in chronic bone infections, although the possibility of antibiotic-associated colitis occurring with this drug should be remembered.

Excretion

The pathway of excretion of an antimicrobial agent should be noted, e.g. drugs metabolized in the liver such as erythromycin estolate should not be given to patients with a history of liver disease as the drug may cause hepatotoxicity leading to acute illness with jaundice.

Toxicity

Most antimicrobials have side-effects and the dentist should be aware of these (see below).

Drug interactions

These are becoming increasingly common due to the extensive use of drugs. For instance, it is now known that antibiotics such as penicillin and erythromycin can significantly reduce the efficacy of some oral contraceptives and that antacids can interfere with the action of tetracyclines. Therefore, it is incumbent upon all practioners to be aware of the drug interactions of any antimicrobial they prescribe. The major drug interactions of commonly used antimicrobials in dentistry are described in Table 14.3.

Table 14.3 Some drug interactions of antimicrobials commonly used in dentistry

Drug affected	Drug interacting	Effect
Penicillins	Probenecid	May potentiate the effect of penicillin. Reduced absorption
	Butazolidin	
	Neomycin	
Erythromycin	Theophylline	Increase theophylline levels leading to potential toxicity
Cephalosporins	Gentamicin	Additive effect leading to nephrotoxicty
	Frusemide (Lasix)	Possible increase in nephrotoxicity
Tetracycline	Antacids	Reduced absorption
	Dairy products	
	Oral iron	
	Zinc sulphate	
Metronidazole	Alcohol	'Antabuse' effect
	Disulfiram	
	Phenobarbitone	Reduced effect
	Phenytoin	

Failure of antimicrobial therapy

A variety of factors outlined below may be involved in the failure of antimicrobial therapy and if an infection does not respond to drugs within 48 hours then prompt consideration should be given to the following:

1. Inadequate drainage of pus or débridement.
2. Inappropriateness of the antimicrobial agent, including bacterial resistance to the drug, dosage and drug interactions.
3. Presence of local factors such as foreign bodies which may act as reservoirs of reinfection.
4. Impaired host response, e.g. in immunocompromised patients due to drugs or HIV infection.
5. Poor patient compliance.
6. Possibility of an unusual infection or that the disease has a non-microbial aetiology.
7. Poor blood supply to tissue.

Antimicrobials commonly used in dentistry

A variety of antimicrobial agents are prescribed by dental

practitioners. However, penicillins remain the most useful and widely used antimicrobial agents in dentistry. For instance, a recent survey in Scotland has shown that penicillin is the first choice antibiotic prescribed by an overwhelming majority of dentists for acute infections in dental practice. In the following section the more commonly used penicillins such as penicillin V and amoxycillin are described in some detail while mentioning in brief the other penicillins which appear in the Dental Formulary.

Penicillins

A wide array of penicillins have been synthesized by incorporation of various side-chains to the β-lactam ring (Table 14.4). The spectrum of activity and indications for the use of these penicillins vary widely; details can be found in the texts given at the end of the chapter.

Table 14.4 Types of penicillin

Group	Type of penicillin
Narrow-spectrum	Benzyl penicillin
	Penicillin V
	Procaine penicillin
	Triplopen
Broad-spectrum	Ampicillin
	Amoxycillin
	Esters of ampicillin
Penicillinase-resistant	Methicillin
	Flucloxacillin
Antipseudomonal	Piperacillin
	Mezlocillin

The commonly used penicillins are remarkably non-toxic but all share the common problem of allergy. Minor reactions such as rashes are common while severe reactions, particularly anaphylaxis, although rare, can be fatal. Allergy to one penicillin is shared by all the penicillins and, in general, the drug should not be given to a patient who has had a reaction to any member of this group. Some 10% of those sensitive to penicillin show cross-reactivity to cephalosporins.

Phenoxymethyl penicillin (penicillin V)

This is an acid-resistant penicillin which can be administered orally. It is commonly used by dental practitioners in the treatment of acute oral infections such as dental abscesses, post-extraction infection, pericoronitis and salivary gland infections. However, penicillin V is less active than parenteral benzyl penicillin (penicillin G) due to its erratic absorption from the gastrointestinal tract. Therefore, in serious infections, penicillin V could be used for continuing treatment after one or more loading doses of benzyl penicillin, when clinical response has begun.

Oral penicillin can cause severe reactions in patients who are allergic. Severe anaphylaxis may occur very rarely. Other uncommon reactions include skin rashes and fever. Despite the above, it is one of the cheapest and safest antibiotics available today.

Penicillin V is effective against a majority of α-haemolytic streptococci and penicillinase-negative staphylococci. Aerobic Gram-positive organisms including *Actinomyces, Eubacterium, Bifidobacterium, Peptococcus* and *Peptostreptococcus* spp. are sensitive, together with anaerobic Gram-negative organisms such as some *Bacteroides, Fusobacterium* and *Veillonella* species.

The majority of *Staph. aureus*, particularly those isolated from hospitals, are penicillinase producers and hence resistant to penicillin. A minority of α-haemolytic streptococci, and some *Actinobacillus actinomycetemcomitans* strains implicated in juvenile periodontitis, are resistant.

Benzyl penicillin (penicillin G)

Benzyl penicillin is useful in moderate to severe infections as its parenteral administration results in rapid, high and consistent antibiotic levels in plasma. However, chances of allergy developing are increased by injection. Administration of greater than 20 megaunits per day to an adult leads to neurotoxicity and anaemia.

Depot penicillins

Procaine penicillin given intramuscularly was formerly popular for endocarditis prophylaxis as it is slowly absorbed to give bactericidal blood concentrations, particularly when administered with benzyl penicillin as fortified procaine penicillin.

Triplopen is a mixture of benzyl, benethamine and procaine penicillins. Initially it gives a bactericidal concentration and thereafter remains bacteriostatic for periods up to 2 – 3 days.

The depot penicillins have limited use and could be employed to initiate treatment in acutely ill children in whom oral absorption is likely to be poor.

Broad-spectrum penicillins

Ampicillin and amoxycillin

These antibiotics are very similar to penicillin but are effective against a broader spectrum of organisms including Gram-negative organisms such as *Haemophilus* and *Proteus* species. Amoxycillin is a derivative of ampicillin and has a similar antibacterial spectrum. It is, however, twice as well absorbed when given by mouth, producing higher plasma and tissue concentrations; unlike ampicillin, absorption is not affected by the presence of food in the stomach.

Ampicillin is sometimes used in the 'blind' treatment of dentoalveolar infections when the antibiotic sensitivity patterns of the causative organisms are not known. In dentistry, amoxycillin is the drug of choice for prophylaxis of infective endocarditis in patients undergoing surgical procedures and scaling (see Chapter 12). A short course of high dose amoxycillin (oral) has been shown to be of value in the treatment of dentoalveolar infections (see chapter 6).

One of the drawbacks of amoxycillin is its susceptibility to β-lactamase. However, a new drug, clavulanic acid, can be incorporated with amoxycillin and protects the latter from the activity of β-lactamase. An amoxycillin–clavulanate combination may be useful in treating head and neck infections when β-lactamase-producing strains of *Bacteroides* spp. are present.

Ampicillins are associated with a higher incidence of drug rashes than penicillin. They should not be administered to patients with infectious mononucleosis (glandular fever) or lymphocytic leukaemia as the probability of a drug rash is very high in these individuals. The rashes, however, are not related to true penicillin allergy. Nausea and diarrhoea are other frequent side-effects, particularly on prolonged administration. In addition, superinfection and colonization with ampicillin-resistant bacteria, such as coliforms and fungi, may occur. The incidence of diarrhoea is probably less with amoxycillin.

Isoxazolyl penicillins

These are narrow-spectrum antistaphylococcal penicillins which are relatively resistant to β-lactamase produced by *Staphylococcus aureus*. Methicillin was the first drug of this group to be discovered and it has now been supplanted by cloxacillin and flucloxacillin which have superior pharmacokinetics.

The main use of cloxacillin and flucloxacillin is in the treatment of confirmed infections due to β-lactamase-producing *Staphylococcus aureus*. They are safe and non-toxic even when used in high doses.

When these antibiotics were introduced almost all *Staph. aureus* were sensitive to these drugs. However, methicillin-resistant *Staphylococcus aureus* (MRSA) have emerged in several regions of the UK and in other parts of the world. The spread of MRSA in the community is being closely monitored to obviate the spread of a highly drug-resistant organism.

Cephalosporins

Cephalosporins are β-lactam antibiotics and, like isoxazolyl penicillins, are relatively stable to staphylococcal penicillinase, although the degree of stability varies with different cephalosporins. Cephradine, which can be given by mouth, and cephaloridine, which can be given by injection, are in the dental practitioners formulary.

They are broad-spectrum antibiotics and active against Gram-positive and Gram-negative bacteria, although individual agents have differing activity against certain organisms.

Cephalosporins have a few absolute indications for their use. In dentistry, cephalosporins should be resorted to as a second line of defence when culture and antibiotic sensitivity test results indicate the suitability of a cephalosporin.

Some 10% of penicillin-sensitive patients demonstrate cross-sensitivity. Allergic reactions, including urticaria and rashes, and anaphylaxis reactions may be seen. Nephrotoxicity has been reported. Another disadvantage of the cephalosporins is that oral bacteria including streptococci may develop cross resistance both to penicillins and cephalosporins. Hence, cephalosporins are not suitable alternatives for a patient who has recently had penicillin.

Erythromycin

Erythromycin has a similar, although not identical, antibacterial spectrum to that of penicillin and is thus the first choice in dentistry for treating penicillin-allergic patients. It has an added advantage of being active against β-lactamase-producing bacteria. It is not usually used as a first-line drug in oral and dental infections since obligate anaerobes are not particularly sensitive.

Erythromycin has a few serious side-effects. The main disadvantage being that high doses (given for prophylaxis of infective endocarditis) cause nausea. The use of erythromycin estolate for more than 14 days may cause cholestatic jaundice.

Lincomycin and clindamycin

These closely related antibiotics have a very restricted use because of their serious side-effects. Clindamycin, which is more active and better absorbed from the gut, has replaced lincomycin which was the original drug in this group.

Clindamycin is mainly used for serious staphylococcal infections in bones and joints and in infections due to Gram-negative anaerobes in the *Bacteroides* group. Clindamycin is particularly effective in penetrating poorly vascularized tissue, notably bone and connective tissue.

Though rare, the most serious side-effect of clindamycin, which can sometimes be fatal, is pseudomembranous (antibiotic-associated) colitis. This complication may occur with most antibiotics but is most frequently seen with clindamycin and is due to a toxin produced by *Clostridium difficile*, an anaerobe resistant to clindamycin. Allergy to these drugs is extremely rare and hypersensitivity to penicillin is not shared by them.

Tetracycline

The tetracyclines are broad-spectrum antibiotics whose usefulness has decreased as a result of increasing bacterial resistance. They remain, however, the treatment of choice for infections caused by intracellular organisms such as chlamydia, rickettsia and mycoplasma as they penetrate macrophages well. A range of tetracyclines is available, although tetracycline itself remains the most useful for dental purposes. The drug is widely distributed in body tissues, becomes incorporated in bone and is particularly concentrated in gingival fluid. Absorption of oral tetracycline is decreased by antacids, calcium, iron and magnesium salts.

In dentistry, tetracyclines have been used with some success as adjunctive treatment in localized juvenile periodontitis. *In vitro* studies have shown that they are effective against many organisms associated with chronic and juvenile periodontitis. In addition tetracycline mouth washes are sometimes useful in alleviating secondary bacterial infection associated with extensive oral ulceration.

Due to the deposition of tetracycline within developing teeth, it should be avoided in children up to 8 years of age and in pregnant or lactating women, otherwise unsightly tooth staining can occur. Diarrhoea and nausea are not infrequent after oral administration due to disturbance to bowel flora. However, when reduced dosages are used, even for prolonged periods, in the treatment of acne and juvenile periodontitis, apparently few side-effects are seen. Serious hepatotoxicity may occur with excessive intravenous dosage.

Tetracycline has a wide spectrum of activity against oral flora including *Actinomyces*, *Bacteroides* spp., *Propionibacterium*, *Actinobacillus*, *Eubacterium* and *Peptococcus* species.

Metronidazole

Metronidazole is an antimicrobial effective against strict anaerobes and some protozoa. It is widely distributed and passes readily into most tissues, including abscesses, and crosses the blood–brain barrier into cerebrospinal fluid. The drug is metabolized in the liver.

It is active against *Bacteroides* spp., fusobacteria, eubacteria, peptococci, and clostridial species. In the UK it is the drug of choice in the treatment of acute ulcerative gingivitis. It is also used, either alone or in combination with penicillin, in the management of dentoalveolar infections. Its main advantages are its very narrow spectrum of activity and low allergenicity.

Minor side-effects of metronidazole include gastrointestinal upset, transient rashes, furred tongue and unpleasant taste in the mouth. Metronidazole interferes with alcohol metabolism and, if taken with alcohol, may cause severe nausea, flushing and palpitations ('Antabuse effect'). Further, it potentiates the effect of anticoagulants and, if used for more than a week, peripheral neuropathy may develop, notably in patients with liver disease.

The sulphonamides and cotrimoxazole

A major advantage of sulphonamides is their ability to penetrate into the cerebrospinal fluid. Hence, sulphadimidine can be given prophylactically to prevent bacterial meningitis in patients who have had severe maxillofacial injuries.

Cotrimoxazole (Septrin) is a combination of a sulphonamide (sulphamethoxazole) and trimethoprim. It has a broad spectrum of activity and can be used in dentistry where there are specific indications, as shown by bacteriological sensitivity tests.

Fusidic acid

This is a narrow spectrum antibiotic with main activity against Gram-positive bacteria, particularly *Staphylococcus aureus*.

Angular cheilitis associated with *Straphylococcus aureus* is a specific indication for the use of fusidic acid in the form of a topical cream. A small percentage of *Staph. aureus* shows resistance to fusidic acid.

Antifungal agents

Antifungal agents commonly used in dentistry belong to the polyenes and the imidazoles. Nystatin and amphotericin are polyene derivatives; miconazole and ketoconazole are two examples of a variety of imidazole antifungals available today (Table 14.5).

Table 14.5 Antifungal agents

Group	Antifungal agent
Polyenes	Nystatin*
	Amphotericin B*
Imidazoles	Miconazole *
	Fluconazole*
	Ketoconazole
DNA-analogues	5-fluorocytosine

* Used in dentistry.

Nystatin
Nystatin is available in the form of pastilles, ready mixed suspension and ointment for the treatment of oral candidoses. This drug binds to the cytoplasmic membrane of the yeast, altering its permeability with resultant leakage of cell contents and death. In very low doses it may behave as a fungistatic agent.

Since almost all freshly isolated strains of oral *Candida* species are sensitive to nystatin it is widely used in the treatment of oral candidoses. An added advantage of nystatin as a topical drug is its extremely low absorption rate from the intestinal tract, mucous membranes and skin. Patient compliance is superior with the flavoured pastille formulation as opposed to the bitter tasting oral suspension or lozenge. Nausea, vomiting and diarrhoea are rare side-effects following ingestion. However, no adverse effects have been reported via the topical route.

Amphotericin B
This belongs to the polyene group of antifungals and is used essentially in the same way as nystatin. The available formulations are lozenges, ointment and oral suspension. As with nystatin, its absorption from the gut is minimal on topical administration. In addition, amphotericin can be administered intravenously for the treatment of systemic candidoses.

Miconazole
This is an imidazole which is available as an oral gel or cream. In addition to its fungicidal effect it is bacteriostatic for some Gram-positive cocci including *Staphylococcus aureus*. This dual action against yeast and staphylococci is useful in the treatment of angular cheilitis. This drug, as with other imidazoles, acts by interfering with the synthesis of chemicals needed to form the plasma membrane of fungi, resulting in leakage of cell contents and death. Resistance only very rarely occurs.

Antiviral agents

In contrast to the multiplicity of antibacterial agents available today, few antiviral drugs have been developed which have proven clinical efficacy. Idoxuridine and acyclovir are in the dental practitioners formulary and may be prescribed for herpetic infections of the oral and perioral regions. However, idoxuridine is infrequently used at present, due to the popularity of the less toxic and more potent new antiviral acyclovir.

Acyclovir
Acyclovir is a recently introduced, efficient antiviral agent; it has a highly selective action and is less toxic than idoxuridine. It is useful in the treatment of primary as well as secondary herpetic stomatitis and herpes labialis.

It has a complex mode of action whereby viral DNA production is blocked at a concentration of some thousand times less than that which is required to inhibit host cell DNA production.

Topical acyclovir (5% cream) can be prescribed by dentists

for herpes labialis. Treatment must be started in the prodromal phase (when there are local tingling or burning sensations) by applying acyclovir cream to the area and repeating it 4 hourly for 5 days. Application at later stages in the cycle of the lesion will correspondingly reduce the time span and the discomfort of the infection.

For treatment of severe herpetic stomatitis or herpes zoster, acyclovir tablets or oral suspensions could be given. The value of the latter used topically in the mouth has yet to be confirmed.

Further reading

Anonymous (1971) Penicillin and the mouth flora. *British Medical Journal*, **2**, 63–64

Cawson, R. A. and Spector, R. G. (1982) *Clinical Pharamacology in Dentistry*, 3rd edn, Churchill Livingstone, Edinburgh

Garrod, L. P., Lambert, H. P. and O'Grady, F. (1981) *Antibiotic and Chemotherapy*, 5th edn, Churchill Livingstone, Edinburgh

Newman, M. G. and Goodman, A. D. (1984) *Guide to Antibiotics in Dental Practice*, Quintessence, Chicago

Part 5
Cross-infection in dentistry

Chapter 15
Viral hepatitis

A clear understanding of viral hepatitis, especially that caused by hepatitis B virus, is essential for all dental practitioners, particularly in view of the serious sequelae of the disease and the potential risk of transmitting the infection to patients by the dentist and vice versa.

Viral hepatitis, in common with AIDS, is one of the few bloodborne viral hazards the dentist has to contend with in his practice. Although the AIDS epidemic has aroused the greatest concern regarding disease transmission, hepatitis B infection can be a much more real danger to both the patient and the dentist, due to the relatively high infectivity of hepatitis B virus compared with HIV. Furthermore, an unknown, but none the less significant number of dentists in the Western world have either contracted viral hepatitis from, or disseminated the infection to patients. In contrast, to date there is only a single documented instance where a dentist is believed to have become infected with HIV during dental treatment.

The term 'viral hepatitis' is commonly used for several clinically similar diseases which cause inflammation of the liver but are epidemiologically and aetiologically distinct (Table 15.1). Two of these, hepatitis A (formerly called infectious hepatitis) and hepatitis B (formerly called serum hepatitis) are well

Table 15.1 Aetiological agents of viral hepatitis

Hepatitis A virus
Hepatitis B virus
Non-A non-B viruses
 (3 types are known)
Delta virus
Herpes simplex virus ⎫
Epstein–Barr virus ⎬ Herpes group
Cytomegalovirus ⎭
Coxsackie B virus
Yellow fever virus

Table 15.2 Epidemiological and clinical features of three types of hepatitis viruses

	Hepatitis B	Non-A non-B hepatitis	Delta hepatitis
Synonym	Serum hepatitis	Hepatitis C	Hepatitis D
Type of virus	DNA	At least 3 different viruses	A defective RNA virus
Incubation period	2–6 months	Variable	Unknown
Transmission	Predominantly parenteral	Parenteral and faecal–oral	Parenteral
Age	All ages	All ages	All ages
Carrier state	Yes	Probable	Yes
Diagnostic tests	Available	Diagnosis by exclusion	Available
Immunity:			
passive immunization	Hyperimmuneglobulin	None	Hyperimmuneglobulin
active immunization	Vaccine (Hepatitis B)	None	Vaccine (Hepatitis B)

characterized, while a third, non-A non-B hepatitis, is thought to be caused by at least three different agents. These viruses, together with the relatively recently discovered delta virus, are of concern in dentistry as they are responsible for the great majority of viral liver diseases. Table 15.2 shows the main differences between the hepatitis caused by hepatitis B, non-A non-B and delta viruses. It is worth remembering that hepatitis caused by the above agents are clinically indistinguishable, although they possess other differences.

The major emphasis of this chapter will be on hepatitis B. However, hepatitis A is discussed below and non-A non-B and delta hepatitis are briefly described at the end of the chapter to outline their relevance to dentistry.

Hepatitis A

The viral agent of hepatitis A (also called infective hepatitis or short incubation hepatitis) is a small, 27 nm, RNA virus that belongs to the picornavirus group, which also includes poliomyelitis and coxsackie viruses. The virus is inactivated by ultraviolet light, exposure to water at 100 °C for 5 min and by exposure to 2% glutaraldehyde for 15 min.

Hepatitis A occurs particularly in parts of the world where sewage disposal measures and food hygiene are unsatisfactory. Although the transmission of the virus may rarely occur via contaminated needles, it is usually contracted by the faecal–oral route through contaminated food and water. Children and young people are most often infected and for this reason a history of hepatitis in childhood would, in most instances, be indicative of a hepatitis A infection rather than hepatitis B.

The diagnosis of acute hepatitis A is confirmed by IgM class antibodies in serum collected during the acute or early convalescent phase of disease. IgG class antibodies, which appear later in the disease, confer enduring protection against the disease. Unlike hepatitis B, there is no carrier state associated with the disease. This, together with the fact that the disease is transmitted faeco-orally, implies that hepatitis A transmission in the dental clinic is highly unlikely.

Passive immunization by hyperimmuneglobulin is effective against clinical illness, particularly when administered in the early incubation period. However, the main use of short-term, pre-exposure prophylaxis is for travellers to hepatitis A endemic areas, such as some parts of the Third World. A successful vaccine for hepatitis A has not yet been developed.

Hepatitis B

Hepatitis B infection is caused by the hepatitis B virus (HBV), a DNA hepadnavirus which is structurally and immunologically complex. Electron microscopy of HBV reveals three distinct particles, 42 nm Dane particle (the complete infective virus), 22 nm spheres, and tubular forms 22 nm in diameter and 100 nm in length (Figure. 15.1).

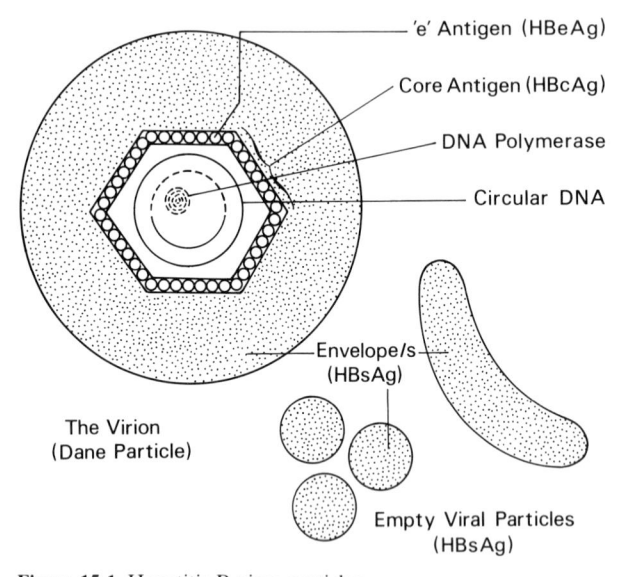

Figure 15.1 Hepatitis B virus particles

The surface of the hepatitis B virus consists of a distinct protein antigen, hepatitis B surface antigen (HBsAg). Overproduction of this surface component within the liver cell cytoplasm generates the non-infectious smaller spheres and tubules which circulate freely in the serum for prolonged periods after the acute episode. Indeed, it is these non-infectious HBsAg particles, abundantly present in plasma of infected individuals, which are the effective component of the plasma-derived hepatitis B vaccine.

The central core of the hepatitis B virus consists of single-stranded DNA, an enzyme (DNA polymerase) and a core antigen (HBcAg). Although this antigen is rarely found in serum, a breakdown product of HBcAg, termed 'e' antigen (HBeAg), may be found in serum and is a marker of active infection. The terminology of hepatitis B virus-associated serological markers is given in Table 15.3.

Table 15.3 Terminology of hepatitis B virus-associated serological markers

Name	Abbreviation
Hepatitis B Virus	HBV
Antigens:	
hepatitis B surface antigen	HBsAG
hepatitis B core antigen	HBcAg
hepatitis B 'e' antigen	HBeAg
DNA polymerase	DNA-P
Antibodies:	
antibody to HBsAg	Anti-HBs
antibody to HBcAg	Anti-HBc
antibody to HBeAg	Anti-HBe

Epidemiology of hepatitis B

Hepatitis B virus is usually spread by the parenteral route rather than the faeco-oral route. The prevalence of hepatitis B varies greatly between different parts of the world, a higher incidence being reported in African and Asian countries than in the Americas, Australia and western Europe. There is a tendency toward higher prevalence in urban than in rural areas, and in men than in women. However, in developed countries the risk of exposure to hepatitis B appears to be greatest in certain categories of people, as shown in Table 15.4.

Table 15.4 Hepatitis B high-risk population groups

Selected patient groups
Patients requiring frequent large volume transfusions of *unscreened* blood/blood products (e.g. haemophiliacs)
Patients in institutions for mentally handicapped
Patients with a recent history of jaundice
Patients in renal dialysis units
Immunosuppressed/immunodeficient groups

Population groups
Intravenous drug abusers
Promiscuous homosexual males
Female prostitutes
Immigrants from Third World, such as Indo-Chinese
Health care and laboratory personnel (especially surgeons)

Carrier state and identification of carriers

Most patients who contract hepatitis B recover within a few weeks without any sequelae (Figure 15.2). However, serological markers of previous HBV infection are invariably present in these patients for prolonged periods. Such markers take the form of antibodies to various components of the HBV.

```
                        Exposure to HBV
                              │
        ┌─────────────────────┼─────────────────────┐
        ▼                     ▼                     ▼
   Subclinical           No effect             Clinical
   acute hepatitis       (? Due to prior       acute hepatitis
                         immunity)
        │                     │                     │
   ┌────┴────┐                │                ┌────┴─────┐
   ▼         ▼                ▼                ▼          ▼
Recovery  Chronic persistent  Chronic     Acute        Recovery
(98%)     hepatitis           active      fulminating  (98%)
          (healthy carrier)   hepatitis   hepatitis
                                  │
                                  ▼
                            Liver damage
                            (cirrhosis)
                                  │
                         ┌────────┴────────┐
                         ▼                 ▼
                   Hepatocellular ──────► Death
                   carcinoma
```

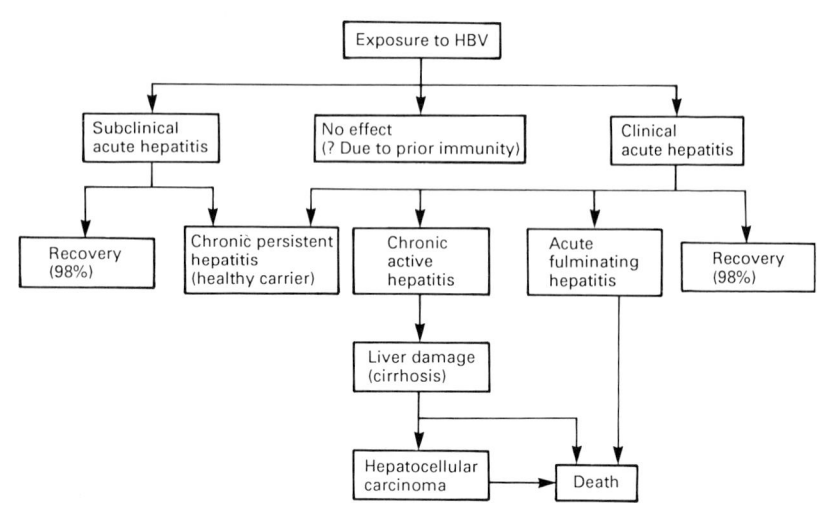

Figure 15.2 Possible sequelae of exposure to hepatitis B virus. Figures in brackets indicate percentage recovery. (From Samaranayake, 1986a, by permission)

A minority (4–5%) of patients, however, fail to clear HBV by 6–9 months and consequently develop a chronic carrier state. It is interesting to note that the latter state more frequently follows anicteric HBV infection (i.e. infection without jaundice). The converse of this is that a majority of infections which lead to jaundice resolve without a carrier state. Hence, a history of jaundice in a patient may indicate, in most instances, little or no risk in terms of hepatitis B transmission.

The chronic carriers of hepatitis B infection fall into two main groups, namely chronic persistent hepatitis (the so-called healthy carrier state) and chronic active hepatitis (Figure 15.2).

In chronic persistent hepatitis the patient does not develop liver damage and he is generally in good health, although the liver cells persistently produce viral antigen (HBsAg) due to the integration of the viral genome into the DNA of the hepatocytes.

The second group of chronic carriers is very infectious as they harbour the infective Dane particles in their blood. In addition, they are very susceptible to cirrhosis and hepatocellular carcinoma. It should be borne in mind that the chronic active hepatitis group represents a very small minority of hepatitis B patients.

In general, therefore, infection with HBV leads to complete recovery in most individuals while a small minority (2–5%) develops a carrier state. These two disease states elicit characteristic serological profiles in the affected individual during various phases of the disease, as shown in Figures 15.3, 15.4.

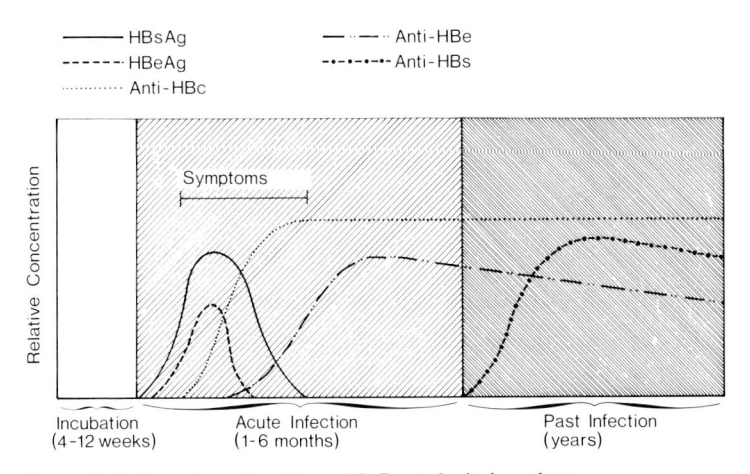

Figure 15.3 Typical profile of hepatitis B serological markers

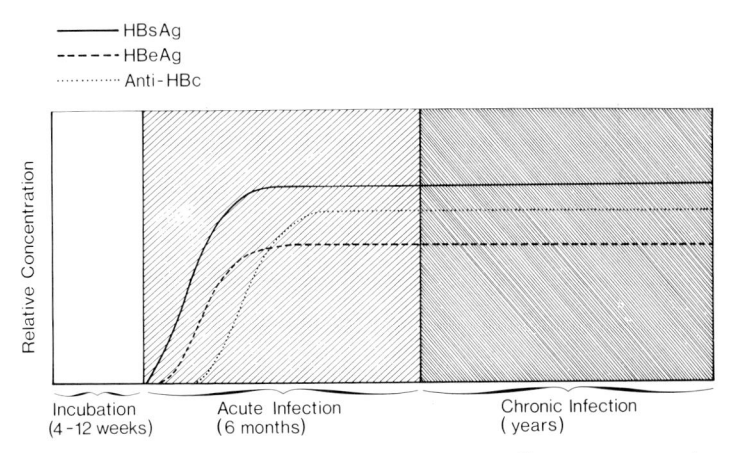

Figure 15.4 Hepatitis B chronic carrier serological profile: no seroconversion

Serological markers and microbiological diagnosis

An understanding of the serological markers of HBV infection is essential in order to comprehend the disease process and to interpret the serological findings. The following is an outline of the important markers associated with HBV infection.

Hepatitis B surface antigen (HBsAg)
The presence of HBsAg indicates that the person is a carrier and potentially infective. This state can persist for months until recovery, or for years in chronic carrier states.

Antibody to hepatitis B surface antigen (Anti-HBs)
This antibody appears in serum during the recovery phase and is a long-lived antibody. The presence of anti-HBs therefore indicates recovery and immunity to further HBV infection. Anti-HBs is seen in high titre after successful vaccination for HBV, as the active ingredient of the hepatitis B vaccine is HBsAg.

Hepatitis B, 'e' antigen (HBeAg)
HBeAg, which is usually found with HBsAg, is indicative of active disease or high infectivity. Infectivity of this particle is so high that even 0.0001 ml serum containing infective virus could transmit the disease to another individual. Chronic persistence of the 'e' antigen in serum would also indicate that the patient will eventually develop chronic liver damage.

Antibody to hepatitis B, 'e' antigen (Anti-HBe)
This antibody appears in the serum soon after the appearance of HBeAg in most patients. Its presence indicates partial recovery from infection and a low level of infectivity while its absence indicates high infectivity and possible chronic carrier state.

The core antigen (HBcAg)
This antigen is present in the liver but not in the serum.

Antibody to hepatitis B core antigen (Anti-HBc)
Anti-HBc in serum is indicative of active or very recent infection. It is a sensitive indicator of previous exposure to HBV infection as it outlasts all other antibodies.

Interpretation of serology
Interpretation of serological results in HBV infection is compli-

cated by the variety of serological markers available to assess the degree of disease activity and the complex sequelae of the disease itself. In Table 15.5 an attempt has been made to simplify the serological interpretation in terms of the hepatitis markers routinely investigated in microbiological laboratories.

Table 15.5 Serological markers of hepatitis B infection and their interpretation

HBsAg	HBeAg	Anti-HBe	Anti-HBs	Risk status
+	Unknown	Unknown	−	High/low risk
+	+	−	−	High risk
+	−	+	−	Low risk
−	−	+	+	Immune due to previous infection
−	−	−	+	Immune due to vaccination

See also Figure 15.3.

Hepatitis B and the dental professional

Hepatitis B in dentistry has been the subject of considerable study. Dental surgeons not infrequently accidentally stab themselves with sharp, infected instruments during routine work and modern equipment exposes dental personnel to potentially infectious aerosols contaminated with blood products and saliva from the patient.

Studies from Denmark and the USA have shown an increased risk of hepatitis B in persons engaged in general dentistry. However, other data from the UK, Norway and Denmark have been unable to demonstrate such an increased risk among dental professionals. Nevertheless, there is general agreement that the risk of contracting hepatitis B is substantially greater among oral surgeons and periodontists than among general dental practitioners. Thus, one US study indicates that general dentists may have a risk of hepatitis thrice that of the general population and surgical specialists run a risk of about six times that of the US population. Furthermore, it is well documented that in all groups the risk of contracting hepatitis B increases with age and with time in practice.

In terms of overt infection it has been estimated that about 5% of dentists in the UK are exposed to HBV at some stage. The average annual attack rate for dentists employed in the

National Health Service (per 100 000) is 17, a figure similar to that for surgeons and physicians.

The usual mode of transmission of hepatitis B is from the patient to the dentist. However, there are at least eight recorded outbreaks where dentists have transmitted the disease to patients. Most of these reports document multiple episodes of hepatitis B transmission directly by an HBeAg-positive dentist, rather than indirect transmission via contaminated instruments. It would also appear that dentists transmitted the disease to the patients from cuts and lacerations on their hands, thus giving rise to a haemo-oral route of transmission. The evidence implicating such transmission is circumstantial and, indeed, some believe that transmission of hepatitis B to patients by dentists is a rare occurrence.

Intraorally, the greatest concentration of hepatitis B virus is at the gingival sulcus, due to the continuous serum exudate which is small in health but greatly increased in diseased states. It is not surprising to find, therefore, that among dental auxiliary personnel, dental hygienists who work primarily in the area of the gingival sulcus are exposed to an equivalent risk to that of the dentist. The virus is not present in parotid saliva.

Identification and dental treatment of patients with a history of hepatitis B

Certain social groups have a high incidence of hepatitis B or antigen carriage (see Table 15.4). A high index of suspicion should be maintained when questioning patients belonging to such groups (e.g. homosexuals and drug addicts) as they do not freely volunteer information. However, only a small minority of those with a history of juandice will have had hepatitis B and even fewer will be carriers.

Nevertheless, patients should be questioned about any past attack of jaundice and the circumstances which led to the illness. An attack of jaundice, particularly in a member of any of the high-risk groups, during adolescence or in adult life and which was associated with joint pains and a rash is strongly suggestive of hepatitis B. Such patients should be serologically tested to assess their infectivity.

It is important to bear in mind that the incidence of HBsAg carriage is increased among the high-risk groups who have never had clinical infection. As it is practically impossible to screen all patients, good and effective precautionary measures (see Chapter 17) should be taken by the practitioner when treating any patient.

The serological status of the patient with a history of hepatitis can be established by contacting the general medical practitioner, area hepatitis reference laboratory or appropriate hospital consultant in microbiology/virology.

If serological tests have not been performed the patient should be referred to his medical practitioner for evaluation of his current serological status. However, if emergency treatment is required, this could be performed by taking the special precautions laid down for high-risk patients (see Chapter 17 and Appendix 2). The serology will indicate whether the patient offers no risk or whether he is a high- or low-risk carrier (Table 15.5). The reader is referred to the Appendices for precautions related to treatment of carriers. If the dentist and the staff have been successfully vaccinated against hepatitis B, then routine dental practice procedures for controlling cross-infection should be more than adequate for treating hepatitis B patients.

Precautions relating to treatment of HBsAg carriers
Clinical procedures, such as adequate cleansing and sterilization of instruments and wearing of protective clothing that minimize exposure to infectious material, must be routinely employed by all dentists. The risk from infectious patients will be minimized if such procedures are routinely implemented. Nevertheless, certain additional precautions are appropriate in the dental treatment of patients carrying blood-borne viruses, including hepatitis B and HIV (Chapter 17).

Antigen-positive dentist and staff

The current position regarding dentists and ancillary staff who are hepatitis B carriers (either high or low risk) is that they may continue in clinical work provided that adequate precautions are taken (the routine use of rubber gloves being essential) and that there is no evidence of transmission of the disease to their patients.

In the unlikely event of dental staff having to cease their employment, they would be entitled to compensation under the National Health Service Injuries Benefits Regulations of 1974.

Hepatitis B prophylaxis

The importance of routine measures such as the taking of adequate medical histories, wearing protective clothing, and proper sterilization of instruments, in preventing cross-infection in the dental clinic is emphasized in Chapter 17. Nevertheless, an additional protective measure for dental personnel who are at risk is hepatitis B vaccine, which confers active immunity on the recipient.

Passive immunization with hyperimmuneglobulin
Protection in the form of passive immunization is available if an unprotected dentist accidentally inoculates himself with blood or saliva during treatment of a hepatitis B risk patient. In such an event, a single injection of hepatitis B hyperimmuneglobulin given within 48 hours of the accident usually aborts or reduces the severity of the infection. Ideally, simultaneous active immunization with the vaccine at a different site should be combined with passive immunization. It has been shown that such combined vaccination does not interfere with the immune response.

Active immunization with hepatitis B vaccines
The active ingredient of the hepatitis B vaccine is the surface protein of the hepatitis B virus, i.e. HBsAg. The antigen can be obtained for vaccine manufacture either from the pooled plasma of infected individuals or by biological or chemical synthetic techniques.

Plasma-derived hepatitis B vaccine This was the first commercially produced hepatitis B vaccine; it is now being supplanted by synthetic vaccines. It consists of alum-adsorbed, inactivated HBsAg particles obtained from pooled plasma of HBsAg carriers. All dentists and auxiliary personnel can obtain the vaccine via the National Health Service by simply paying the prescription charge. The vaccine is administered as three separate doses. The first is given at an elected date, the second, one month later and the third, six months after the first dose.

It is generally well tolerated with minimal side-effects. The rate of seroconversion in healthy vaccine recipients approaches 95%. As some 5% may not seroconvert it is important for antibody levels to be measured after a full course of vaccination. Those who fail to respond are usually given another inoculation of the vaccine. It is becoming clear that there is a wide individual variation in the titre of antibody produced after a full course of vaccine. As a result it is likely that although some vaccinees will not require a booster for 5 years, others may require the vaccine much earlier.

Biologically and chemically synthesized hepatitis B vaccines The plasma-derived hepatitis B vaccine has several limitations, not the least being its high cost. Further, production is dependent on the availability of infected human serum, which requires elaborate purification and inactivation procedures. Hence, scientists have successfully developed biologically and chemically synthesized vaccines which are now supplanting the first generation plasma-derived vaccines. Vaccines produced by the yeast recombinant DNA technique are currently available in the UK and their manufacture is outlined below.

In brief, the gene which codes for the HBsAg is isolated from the hepatitis B viral genome and is inserted into a plasmid vector (intermediary vehicle). The plasmid, now containing the HBsAg gene, is in turn introduced into the secondary host, which is the yeast, *Saccharomyces cerevisiae*. Clones of the secondary host can be obtained to produce high yields of HBsAg which are structurally and therefore antigenically similar to human HBsAg. The yeast recombinant hepatitis B vaccine thus produced is as effective and safe as the first generation, plasma-derived vaccine.

Inoculation injury involving a hepatitis B risk patient

In the event of an inoculation injury involving a hepatitis B risk patient the following procedures should be followed.

1. When a needle-stick injury occurs, ensure that the accident is not repeated by discarding needle/sharp into an appropriate container. (Resheathing needles without a suitable 'needle safety device' is dangerous and should be avoided.)
2. Encourage wound to bleed freely and wash thoroughly with a proprietary chemical hand-wash (e.g. Hibisol) and hot water.
3. If the patient falls into a high- or low-risk category and the dentist is not immune, hyperimmune antisera should be administered as soon as possible and not less than 48 hours after the accident. Arrangements should also be made to receive a course of hepatitis B vaccine, the first dose of which should preferably be administered with the hyperimmuneglobulin, but at a different body site.
4. If the patient falls into a risk category and the dentist is immune, no further action is necessary.
5. If the patient's serological status is unknown and the dentist is not immune, prior vaccination with hepatitis B vaccine is

the most effective method of preventing disease transmission because
(a) Some health authorities do not administer hyperimmune sera to injured personnel unless the patient poses a definite risk of transmitting hepatitis B.
(b) It may be difficult to establish the serological status of the patient within 48 hours.
6. If the dentist has received either a part or a full course of the hepatitis B vaccine prior to the accident, it may be reassuring for him to establish his own antibody levels.

Non-A non-B (NANB) hepatitis

In the last few years many cases of NANB hepatitis have been recognized. This disease, sometimes called hepatitis C, occurs most commonly after blood transfusion and parenteral drug abuse. Diagnosis is based on the exclusion of the other known aetiological agents of hepatitis by serological tests (see Table 15.1).

Worldwide, NANB hepatitis seems to represent 15–25% of clinical cases of viral hepatitis, although it seems more common in Japan. Three epidemiological types of NANB hepatitis have been recognized; the epidemic water-borne type; the blood- or needle-associated type; and the sporadically occurring (community-acquired) type. Therefore, transmission of NANB viruses could occur via contaminated needles, food or water.

A universally recognized, specific serological test has not yet been developed for NANB virus infections. At present, there are no active or passive immunoprophylaxis measures available for persons with percutaneous exposure to blood from a patient with NANB hepatitis. However, some authorities believe that it may be reasonable to administer hyperimmuneglobulin immediately after accidental exposure to NANB infections.

Delta hepatitis (hepatitis D)

Delta hepatitis is caused by a recently discovered 'defective' RNA virus which coexists with the hepatitis B virus (Figure 15.5). Delta infection may therefore be seen as either a coinfection in a hepatitis B patient or a superinfection in a hepatitis B carrier, each of which usually causes an episode of acute hepatitis. Coinfection usually resolves, while superinfection frequently causes chronic delta infection leading to chronic active hepatitis. Areas with a high incidence of delta hepatitis are the Amazon basin, parts of Africa, the Middle East and Arab countries where 30–90% HBsAg carriers with liver disease are infected. Delta infection occurs rarely in the susceptible population of northern Europe and is virtually confined to parenteral drug abusers.

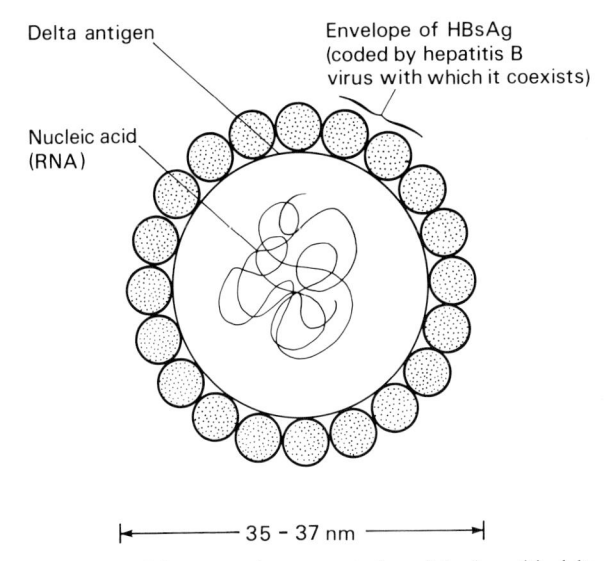

Figure 15.5 Diagrammatic representation of the hepatitis delta virus

Delta hepatitis may be diagnosed by detection of delta antigen in serum and/or by the appearance of delta antibody. A test for detection of the delta antibody is commercially available.

Routes of delta transmission appear to be similar to those of hepatitis B, the infection being most commonly seen among persons at high risk of acquiring hepatitis B infection (see Table 15.4). It is particularly interesting to note that, as the delta virus is dependent on hepatitis B virus for replication, immunization with the hepatitis B vaccine will suffice to prevent delta infection in at-risk groups.

Further reading

Follett, E. A. C. and MacFarlane, T. W. (1981) Infectivity of Hepatitis B surface antigen (Australia antigen) positive patients. *British Dental Journal*, **150**, 92–93

Mori, M. (1984) Status of viral hepatitis in the world community: its incidence among dentists and other dental personnel. *International Dental Journal*, **34**, 115–121

Samaranayake, L. P. (1986a) Viral hepatitis: 1. Aetiology, epidemiology and implications. *Dental Update*, **23**, 353–365

Samaranayake, L. P. (1986b) Viral hepatitis: 2. Hepatitis B vaccines. *Dental Update*, **13**, 411–416

Samaranayake, L. P., Scully, C., Dowell, T. B. et al. (1988) New data on the acceptance of the hepatitis B vaccine by dental personnel in the United Kingdom. *British Dental Journal*, **164**, 74–77

Scully, C. (1985) Hepatitis B: an update in relation to dentistry. *British Dental Journal*, **159**, 321–328

Chapter 16
Human immunodeficiency virus and acquired immune deficiency syndrome

Infection with human immunodeficiency virus (HIV) is probably the most lethal infection which can be contracted via parenteral routes. This is because the scientific community has not yet developed an antiviral agent or a vaccine which successfully combats the virus. Although the theoretical possibility of contracting hepatitis B in the dental clinic is far greater than that of acquired immune deficiency syndrome (AIDS), this is counterbalanced by the very high mortality and, indeed, the stigma and social consequences associated with the latter. These, together with the extensive media coverage and intense public awareness of AIDS, mean that the average patient would probably be more concerned about contracting AIDS than hepatitis B. Therefore, the dental practitioner should be knowledgeable on AIDS, firstly to diagnose patients who may present to him with early signs and symptoms of the disease, secondly to implement appropriate infection control measures and finally to educate, advise and allay the public fears concerning AIDS transmission both within and outwith the dental clinic environment.

Definition of AIDS

Acquired immune deficiency syndrome is the name given to a group of disorders characterized by a profound cell-mediated immunodeficiency consequential to irreversible suppression of T lymphocytes by the human immunodeficiency virus. The affinity of HIV to reside in and destroy T lymphocytes in turn affects the other arms of the immune system, causing generalized susceptibility to infections and neoplastic diseases and ultimately leading to the death of the infected individual.

Epidemiology

Although AIDS is now recognized internationally as a disease of worldwide prevalence, its origins and spread are still a matter of controversy. The first five cases of AIDS were noted by the health authorities in the USA in 1981, while its transatlantic spread to the UK was noted a year later. Since then there has been an exponential rise in the number of AIDS patients in the UK with a doubling of reported cases every 10 to 11 months. At the time of writing, a total of 1730 AIDS patients have been reported in the UK and 96500 AIDS cases have been reported to the World Health Organization from 136 countries. Almost all full blown AIDS patients die within 18 months and the antibody-positive pool of patients worldwide is estimated to be somewhere between one and several hundred thousand.

The main risk groups for AIDS include sexually active male homosexuals and bisexuals, recipients of blood transfusions or products, and intravenous drug abusers. In the African continent men and women are equally affected, as opposed to fewer than 10% of female patients in the UK and USA. AIDS in Africa is spreading rapidly and is often characterized by a wasting syndrome termed 'slim disease'. Paediatric AIDS is seen in a significant proportion of infants born to infected mothers or those who received contaminated transfusions. The high risk groups for AIDS and AIDS-related diseases are shown in Table 16.1.

Table 16.1 Major risk groups for AIDS and related diseases

Sexually promiscuous homosexual or bisexual men
Intravenous drug abusers who share injection equipment
Haemophiliacs and other recipients of *unscreened* blood or blood products
Sexual partners of the above groups or infected persons
Infants born to infected mothers

Human immunodeficiency virus

The major aetiological agent of AIDS is the human immunodeficiency virus. However, it is believed that cofactors such as immunosuppressive drug therapy and the genetic haplotype of an individual may play a contributory role in the pathogenesis of the disease.

Human immunodeficiency virus is an RNA virus belonging to a family of recently discovered retroviruses which have been known to cause leukaemias, lymphomas and immune deficiency conditions in animals. Three types of human retroviruses have been characterized so far: (1) human T cell lymphotropic virus I (HTLV I), which is associated with adult human T cell leukaemia; (2) HTVL II, which was originally isolated from a patient with hairy cell leukaemia but has not been linked to any particular disease; and (3) HIV, which causes AIDS. Whereas HTLV I and II are stable viruses with a low replication rate, HIV has a high replication rate and shows variability in its outer envelope. Indeed, French workers have isolated a variant of HIV from an AIDS patient which they have provisionally designated as HIV II.

The human immunodeficiency virus has a genome of RNA surrounded by two layers of protein (P24 and P18). These layers are in turn covered by a lipoprotein layer which has glycoproteins embedded in it as 'knob-like' structures (Figure 16.1). Along with the RNA are molecules of an enzyme, reverse transcriptase, which the virus uses during multiplication.

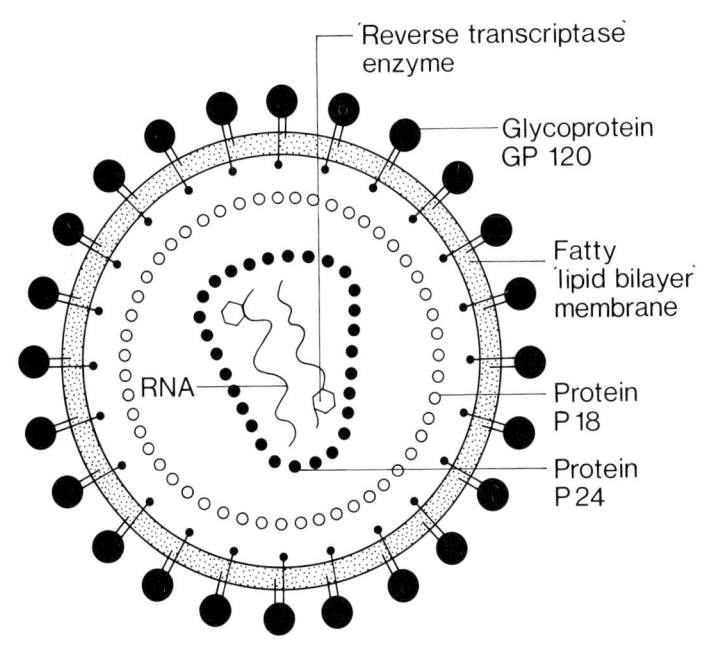

Figure 16.1 A simplified diagram of the human immunodeficiency virus

The virus can enter only the host cells which have specific receptors on their surface. In particular, the virus binds to helper T lymphocytes, known as T4 cells. Once the virus enters the host cell, the reverse transcriptase first copies the viral RNA into DNA (as RNA can not be incorporated into the host DNA). The viral RNA, now masquerading as DNA, then enters the nucleus of the cell where it is integrated into the cell's own DNA, thus becoming a permanent part of an infected person's own cells.

The original infective virus has now become a provirus which resides in a quiescent phase in the cellular DNA until it is activated by an external stimulus. The latter could possibly be another infection or foreign material which stimulates the lymphocytes (T4 cells) and concomitantly activates the proviral gene. Now the cell begins to make copies of the viral RNA in the form of messenger RNA. This travels out of the nucleus to the cytoplasm and initiates production of viral proteins, in addition to making viral RNA. The viral proteins and RNA then assemble into infective viral particles and bud out from beneath the cell membranes of the infected cell. As they do so, they acquire the outermost lipoprotein membrane from the cell wall and are now in a state to infect other lymphocytes. It is thought that the parent T4 cell dies due to leakage of cell constituents resulting from the loss of the cell wall.

As mentioned above, HIV has an exquisite affinity for human T lymphocytes, particularly a subset of lymphocytes, the T helper phenotype (also called OKT4 component) which plays a central role in the inducer/helper function of the immune response in humans. The cell-mediated immune system is most severely affected, both qualitatively and quantitatively. However, the immune defects observed are broad and diverse, reflecting the major influence which the helper T subset of lymphocytes has on the entire spectrum of the immune system. Furthermore, it is now known that HIV is neurotropic and grows readily in brain tissue (neuroglial cells), causing behavioural abnormalities and memory loss, particularly at the late stages of the disease. The virus is also found in many body fluids including blood, saliva, semen, tears and breast milk.

The proof of the involvement of HIV in AIDS has come from the following: (1) the isolation of the virus from patients with AIDS and AIDS-related complex (ARC); (2) the presence of high concentrations of serum antibodies to the virus in these patient groups; (3) the transmission of the virus through blood transfusions; (4) the ability of the virus to cause specific cytopathic effects in T cells; and (5) the ability of the virus to cause AIDS-like diseases in chimpanzees.

Antibodies to the virus (HIV antibody) appear as a response

to the infection. The antibody response, however, does not mean that the patient has recovered and that the virus has been eliminated, as is the case with most viral infections. Therefore, at present all HIV antibody-positive individuals are regarded as potentially infective. A few patients with AIDS have no detectable HIV antibody and hence the absence of the antibody is not a definitive criterion for disease exclusion.

Some 20–60% of individuals who are antibody positive are probably at risk for developing AIDS. Moreover, 90% of patients with full-blown AIDS are seropositive.

Natural history and clinical features of AIDS

AIDS is an insidious disease process with protean manifestations. Consequently a number of stages in the natural history of

Table 16.2 Opportunistic infections, neoplasms and other features of AIDS and related diseases

Opportunistic infections	
Pneumonia, sinusitis:	*Pneumocystis carinii*
	Aspergillosis
	Candidosis
	Cryptococcosis
	Strongyloidosis
	Mycobacterioses
	Legionellosis
Gastrointestinal:	Cryptosporidiosis
	Isospora
	Giardiasis
Central nervous system:	Polyomavirus
	Toxoplasma gondii
	Papovavirus
Mucocutaneous:	Herpesvirus
	Papillomaviruses
	Mycobacteria
	Candidosis
	Histoplasmosis
Neoplasms	
Kaposi's sarcoma	
Lymphoma	
Squamous cell carcinoma	
Miscellaneous complications	
Encephalopathy	
Idiopathic thromobocytopenic purpura	
Lupus erythematosus	
Seborrhoeic dermatitis	

AIDS has been identified, ranging from subclinical infection, persistent generalized lymphadenopathy (PGL), AIDS-related complex, and full-blown AIDS (Figure 16.2). The most common syndrome of AIDS is PGL which is due to reactive hyperplasia of lymph nodes.

The prodromal illness of AIDS is characterized by malaise, low-grade pyrexia, diarrhoea, weight loss and generalized lymphadenopathy. Oral candidosis is frequently a feature.

The opportunistic infections, neoplasms and other features of AIDS and its prodrome are listed in Table 16.2. However, by far the most common of these opportunistic infections, seen in 50% of AIDS patients, is *Pneumocystis carinii* pneumonia, while the most common neoplasm is Kaposi's sarcoma. As HIV is neurotrophic, up to 60% of full-blown AIDS patients may

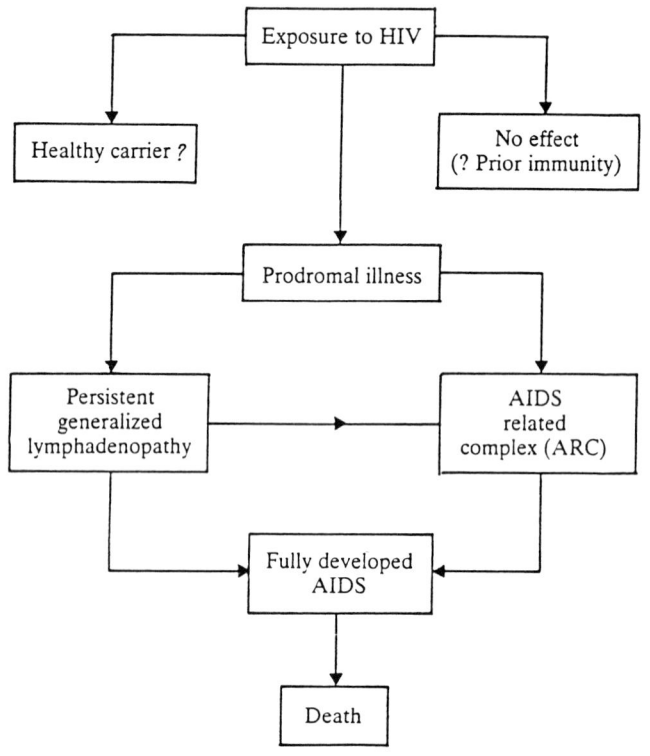

Figure 16.2 Possible sequelae of exposure to human immunodeficiency virus. (From Samaranayake, 1987, *General Dental Treatment*, edited by W. M. Tay, by permission)

have clinical manifestations of encephalopathy, including dementia and paralysis.

Oral manifestations of AIDS

A large majority of AIDS patients have oral manifestations and, indeed, such manifestations are the presenting signs of the disease in some individuals. The relative frequency of oral lesions in AIDS patients is shown in Table 16.3, while the more common lesions are described in some detail below.

Table 16.3 Oral manifestations of AIDS

Fungal infections
  Oral candidosis[a]
    pseudomembranous
    erythematous
    hyperplastic
      plaque-like
      nodular
  Angular cheilitis
  Oral histoplasmosis
Viral infections
  Herpetic stomatitis[a]
  Hairy leukoplakia
  Oral zoster
  Oral condyloma acuminatum
  Cytomegalovirus
    xerostomia
Bacterial infections
  HIV-gingivitis[a]
  HIV-necrotizing gingivitis
  HIV-periodontitis[a]
  Submandibular cellulitis
  Mycobacterium avium intracellulare infection
  Klebsiella pneumoniae
  Enterobacter cloacae
Neoplasms
  Oral Kaposi's sarcoma[a]
  Oral squamous cell carcinoma
  Non-Hodgkin's lymphoma
Unknown aetiology
  Recurrent aphthous ulceration
  Idiopathic thrombocytopenic purpura
    oral ecchymoses
  Salivary gland enlargement

[a] More frequent entities.

Oral candidosis

Oral candidosis (thrush) is a frequent finding in AIDS patients and is thought to present as an intermediate stage in the spectrum of illness that progresses from ARC without candidosis, to oral candidosis and then to AIDS. As oral thrush is present in about 75% of both ARC and AIDS patients, it has a significant predictive value for the subsequent development of full-blown AIDS. Hence, the clinician should be aware of the implication of discovering oral candidosis in an otherwise apparently healthy young individual. Oral candidosis also implies the concurrent presence of oesophageal candidosis and increased likelihood of other systemic opportunistic infections. Oral candidosis in AIDS patients has been classified as outlined in Table 16.3. It is thought that the location and type of candidal lesion is important, as different types of candidosis may be associated with different causes and/or cofactors.

The pseudomembranous variety of oral candidosis associated with AIDS may persist for months, as opposed to the classic, acute pseudomembranous candidosis seen in infants which remits within days after antifungal therapy. It is characterized by creamy white or yellowish plaques on a red or normal-coloured oral mucosa which can be scraped to reveal a raw bleeding surface. The soft and hard palates, buccal and labial mucosa and the tongue are most often affected although it may involve any part of the oral mucosa.

As the name implies the erythematous or atrophic variety of candidosis appears as a red lesion, particularly on the palate and dorsum of the tongue. Focal erythematous lesions may also be seen in the buccal mucosa. These lesions persist for prolonged periods despite adequate antifungal therapy.

Chronic hyperplastic candidosis in AIDS patients is commonly found bilaterally or unilaterally on the buccal mucosa and only rarely on the retrocommissural region. Chronic angular cheilitis caused by *Candida* species and/or *Staphylococcus aureus* may also be seen. Other fungal infections such as disseminated histoplasmosis have been described in a small number of AIDS patients.

Viral infections

Infections with the herpes group of viruses are common in ARC or AIDS patients due to the generalized lowered activity of T cells. The herpes simplex virus (HSV) infections are particularly severe and persistent but rarely disseminate and respond well to acyclovir therapy. However, frequent and troublesome relapses are usually seen as crops of small vesicles on the hard palate or gingiva which become ulcerated and secondarily infected. Varicella zoster infections are less common than HSV infections.

Lesions caused by human papilloma viruses, such as oral papillomas, verruca vulgaris, condylomata and focal epithelial hyperplasia, commonly occur in HIV-infected patients. A number of examples of oral warts have been seen in HIV-positive homosexuals.

Hairy leukoplakia

A new clinical entity designated hairy leukoplakia has been described in HIV-infected patients (Figure 16.3). It is particularly characteristic in homosexual men who develop AIDS and derives its name from the raised white areas of thickening on the lateral border of the tongue. Hairy leukoplakia (HL) can also appear on the dorsum or central surface of the tongue, buccal mucosa and the floor of the mouth and palate. On these surfaces, HL appears as a flat lesion without corrugations. The patients are usually asymptomatic.

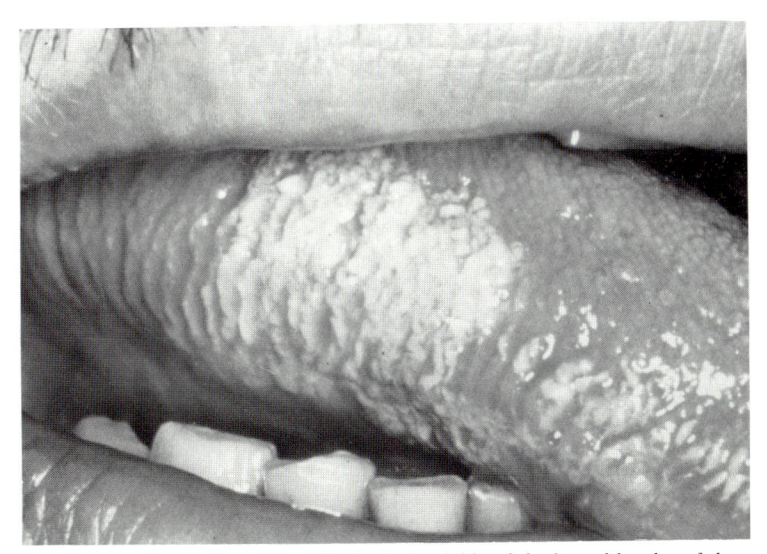

Figure 16.3 Classic appearance of hairy leukoplakia of the lateral border of the tongue in a patient with full blown AIDS

Hairy leukoplakia is now thought to be a viral-induced lesion as the presence Epstein–Barr virus (EBV) has been demonstrated, but this has not yet been conclusively proved, since EBV can be isolated from normal oral tissues. Evidence of candidal infection is found in a significant proportion of lesions, but the lesions do not resolve after antifungal therapy. As preliminary reports indicate that approximately one in three individuals with HL will develop AIDS, its presence may indicate the necessity for further specialist referral. The lesion should nevertheless be differentially diagnosed from other similar clinical entities including idiopathic leukoplakia, lichen planus, chronic hyperplastic candidosis, geographic tongue and occlural trauma.

Opportunistic bacterial infections

Although oral fungal and viral infections are almost invariably seen in HIV-infected patients, opportunistic bacterial infections are not as common. However, gingivitis and periodontitis have been diagnosed in an increasing number of antibody-posistive patients or those with AIDS. HIV-gingivitis presents as either localized or generalized lesions with spontaneous haemorrhage and petechiae in gingival margins. These may progress to nectrotizing ulcerative gingivitis. The term acute ulcerative gingivitis is not used due to the chronicity and recurrence of the disease. In some patients necrotizing gingivitis may take a rather destructive course leading to loss of soft tissue and bone. HIV-periodontitis manifests with deep pain and spontaneous bleeding, interproximal necrosis and cratering, severe erythema and a higher rate of bone loss than in necrotizing gingivitis. In general, gingivitis and periodontitis seen in AIDS are more severe and aggressive than the classical forms of the disease, with the result that relapses are common.

Other bacterial infections which are rarely, if ever, found intraorally have been described in HIV-infected individuals. They include infections caused by mycobacteria, and coliforms.

Lymphadenopathy

Over 95% of AIDS patients who present with lymphadenopathy exhibit cervicofacial lymph gland enlargements. Lymph nodes may be variable with respect to size and consistency and prominent areas involved include posterior cervical, occipital and anterior superior cervical glands.

Oral neoplasms

Several neoplasms can manifest intraorally due to AIDS. Kaposi's sarcoma (KS), an edothelial cell tumour, is the characteristic malignant neoplasm found in AIDS and, in 50% of patients, develops in the oral or perioral area. It is one of the earliest oral manifestations of AIDS and presents as a red or purple macule or a nodule, usually in the palate. The finding of oral KS in a male who is otherwise well is virtually pathognomonic of AIDS.

Oral squamous cell carcinoma and oral lymphomas may develop infrequently in HIV-infected individuals.

Miscellaneous oral manifestations

Other, less frequent oral manifestations of AIDS include osteomyelitis, malignant melanomas, recurrent aphthae and parotitis.

HIV in saliva

As mentioned earlier, HIV has been isolated from saliva; however, it is unlikely that saliva can be considered a major reservoir of infection. For instance, in one study HIV could only be cultured from one of 83 saliva samples from antibody-positive individuals, as opposed to 56% cultured blood samples which yielded HIV. Recently, it has also been shown that the infectivity of HIV is significantly reduced on exposure to saliva.

Almost all the studies on saliva have been carried out with mixed saliva. Obviously mixed saliva contains crevicular fluid and therefore it is likely that the HIV in saliva could have originated from the serum present in crevicular exudate. Further studies with pure parotid or submaxillary saliva are needed to clarify this point. It is noteworthy that hepatitis B virus has been detected in whole saliva but not in pure parotid secretions.

Studies have also been conducted for the presence of salivary antibodies to HIV. In these investigations some 70–90% of patients demonstrated salivary antibodies to certain components of the virus, whereas all of them were positive for serum antibodies. Most of the salivary antibody response has been attributed to IgA class antibodies.

The above data indicate that the presence of HIV in saliva is low, while the corresponding antibody levels may be relatively high. This data, together with the epidemiological data which do not support a major role for a salivary route for AIDS, should be comforting facts for all dental practitioners.

Transmission of AIDS

Transmission of AIDS is primarily through receptive anal intercourse and by administration of blood or blood products, including plasma. As it is thought that a high dose of HIV is required to contract the disease, the number of sexual partners and frequency of sexual activity are significant factors in infectivity. However, susceptibility seems to play a role in transmission, since individuals who are not in high-risk groups but who are exposed to the virus appear to remain seronegative.

In blood, the virus has been isolated from patients with AIDS or AIDS-related complex (see below), and healthy persons having antibodies to HIV.

The transmission of AIDS to health care personnel who knowingly or unknowingly treat HIV-infected patients is being continuously monitored due to the grave implications of contracting this fatal disease. Data emanating from the Centre for Disease Control in Atlanta, USA indicate, at the time of writing, that there have been three health care workers who were exposed to the blood of patients with AIDS who subsequently developed antibodies against HIV. None of them however had needle-stick injuries or any other alternative risk factor to explain their infections and the only way for the virus to have been transferred was thought to have been a clinical setting where the workers skin had come into contact with the blood of HIV antibody-positive patients.

In addition to these three cases, only four further cases of acquired infection in workers, in whom seroconversion occurred as a result of recognized exposure incidents, have been documented. Falling into this category, and of special interest to dental personnel, is a reported single case among 1200 prospectively surveyed dental care workers in the USA who probably acquired HIV infection by accidental exposure. This particular individual had no other compromising risk factor.

None of the above individuals, although antibody positive, have thus far developed full-blown AIDS. On the contrary, in some individuals the antibody levels have been shown to decline from the peak levels seen after the exposure. In summary, with one exception, none of the dental personnel in western Europe and the USA have so far contracted HIV infection from their patients during treatment.

The risk of AIDS transmission via normal social contact, such as shaking hands, or in a domestic setting has been investigated and such studies have failed to demonstrate HIV transmission to the contacts, including family members.

On comparing the infectivity of the HIV and the hepatitis B virus it is well known that the latter is more infectious than the former. Thus, two health care workers who sustained needle-stick injury from AIDS patients have been asymptomatic for HIV up to two years after the accident but contracted hepatitis B as a consequence. Further, it has been estimated that in hepatitis B patients the titre of viral particles in blood is about 100 000 000/ml and carries a 6–30% risk of infection following a needle-stick injury. However, the HIV titre is 100–10 000 viruses/ml blood and carries approximately a 1% risk of transmission.

Another contributory factor concerning the apparently lower infectivity of HIV is that, by the time the person is diagnosed as having AIDS, the potential for transmission of the disease appears to be low. One explanation for this is that the patient has a low concentration of HIV in the peripheral blood due to depletion of lymphocytes, which are the cells where the virus replicates.

The corollary of the above is that dental personnel should be more concerned with the unknown, asymptomatic patient in risk groups who may not have been diagnosed as having AIDS, and who theoretically pose a greater transmission risk as they may harbour a significant viral load within their blood.

To conclude, the available data indicate that dentists run a low risk of contracting AIDS by virtue of their occupation and it would be sensible for all dental personnel to be vigilant in standards of hygiene and safety while carrying out all routine procedures on all patients.

Control of HIV infection

As there is no reliable cure for AIDS, the only method of preventing disease transmission is public eduction in avoidance of sexual promiscuity and high-risk sexual practices such as unprotected intercourse; discouraging needle sharing among drug addicts is also important.

Intense efforts are being made worldwide to develop a vaccine which it is thought will take at least another five years to produce. Difficulty in vaccine production is mainly due to the genetic variability of the viral genome which helps the organism to evade the host–antibody response.

Laboratory diagnosis

Clinical or subclinical infection with HIV can now be ascertained by serological tests which detect the antibody response (anti-HIV) to the infection, as most individuals seroconvert within the first 6–12 weeks. The significance of the anti-HIV titre is, however, unclear as false positive and false negative results are relatively common for a variety of reasons. The presence of antibody does mean that the patient has been exposed to the virus and is probably a carrier; it does not indicate immunity to the infection.

Testing for HIV antigens is being used in a few laboratories to detect the early period of infection which precedes antibody production. It is thought that a sensitive, specific test for HIV antigen detection may replace the currently used antibody tests in the near future.

Management

As a result of the medical history and/or the presentation with the foregoing oral manifestations, the dental practitioner may be alerted to the possibility that a patient may be infected with HIV. However, it is imperative that the clinician should not mention the possibility of AIDS unless the patient asks leading questions about the disease. Because of the catastrophic sequelae and the stigma of AIDS, discussing such a possibility with the patient should be the remit of those who are skilled in counselling. If there are oral signs which give cause to suspect a patient of being an early sufferer, then consultation with the patient's general medical practitioner is mandatory, with subsequent referral to a specialist centre for diagnosis.

There is no effective treatment as yet, either for the immune defect or for HIV infection. The mainstay of treatment is alleviating the opportunistic infections or neoplasms by appropriate therapeutic measures.

New treatment modalities for immune defects, such as thymic hormones, interferon and thymosin, are being examined together with antivirals, including ribavarin and azidothymidine (AZT). The latter appear to show some promise.

Duties of dentists with HIV or suffering from AIDS

There is no evidence to indicate transmission of HIV from dentist to patient in the course of treatment. However, there is considerable public concern about the risk of contracting AIDS and about the possibility that patients might be infected by dentists who are suffering from AIDS. Therefore, it is the ethical and moral responsibility of dentists who believe that they may have been infected with HIV to obtain medical advice and, if found to be infected, to submit to regular medical supervision. This should include counselling in respect of any changes in the practice routine which might be considered appropriate in the best interests of their patients. It is the duty of an infected dentist to act upon this medical advice and, if necessary, to modify the practice in some way or cease the practice of dentistry altogether. HIV-positive dentists who knowingly jeopardize the health of their patients by failing to obtain or act on medical advice may be deemed guilty of serious professional misconduct.

Further reading

Barr, C. E. and Marder, M. A. (1987) *AIDS: A Guide for Dental Practice*, Quintessence, Chicago

Greenspan, D., Pindborg, J., Greenspan, J. S. and Schidt, M. (1986) *AIDS and the Dental Team*, Munksgaard, Copenhagen

Klein, R. S., Harris, C. A., Small, C. B. et al. (1984) Oral candidosis in high-risk patients as the initial manifestation of the acquired immunodeficiency syndrome. *New England Journal of Medicine*, **311**, 354–358

Porter, S. R., Scully, C. and Cawson, R. A. (1986) AIDS: update and guidelines for general dental practice. *Dental Update*, **14**, 9–17

Samaranayake, L. P. and Lewis, M. A. O. (1985) AIDS and the dental practitioner. *Dental Update*, **12**, 551–562

Chapter 17
Cross-infection and sterilization

Cross-infection in the dental surgery

Cross-infection may be defined as the transmission of infectious agents among patients and staff within a hospital environment. Transmission can occur between individuals by direct contact but more commonly it involves fomites, i.e. contaminated inanimate objects. In order to understand cross-infection in the dental surgery it is important to have some idea of the source of the infective agents, the means by which they are transmitted, and the route by which they find their way into the tissues of the recipient. In dentistry the three main sources of infection are: (1) patients suffering from the infectious disease; (2) patients in the prodromal stage of certain infections; and (3) individuals who are carriers of pathogenic microorganisms.

It is rare for a patient who is suffering from an acute infectious disease to require dental treatment, although this may occur in an emergency situation. A more likely source of infection is a patient who attends for treatment during the prodromal stage of an infectious disease. At this stage of the infection the secretions and blood of the patient may be infectious to others, although the patient may appear and feel relatively well. Certain bacterial and viral infections have a prodromal stage, although the phenomenon is more common with viral agents (e.g. mumps, measles and chickenpox) than with bacteria. Another group of patients who may be potentially infectious to their dentist during treatment are those who are carriers of various pathogenic agents. There are two types of carriers: (1) convalescent carriers, who have suffered from infections and have apparently recovered but still shed microorganisms into their blood and certain of their secretions; and (2) asymptomatic carriers, who give no past history of infection. To all intents and purposes it is almost impossible for a dentist to diagnose accurately patients who are in the prodromal stage of an infection or those who are carriers. Therefore it makes sense for a dentist to assume that all patients are potentially infectious and to take reasonable precautions to protect both himself and his staff from infection during treatment. In addition, he must ensure that his patients do not become infected during dental treatment in his surgery.

Transmission of infection

There are three ways in which transmission can occur during dental treatment: (1) by the direct contact of tissues with saliva or blood; (2) by droplets containing infectious agents; and (3) via contaminated instruments which have not been rendered safe for re-use.

Route of infection

Microorganisms enter the tissues of the recipient by means of injection through intact skin or mucosa (usually due to an accident involving a sharp instrument) or via defects in the skin (for example, recent cuts or abrasions). Another possible route of infection is by inhalation of infected droplets.

Therefore, it is clear that, during treatment in a dental surgery, pathogenic microorganisms could be transmitted both to and from patients and staff in a number of ways. There is little doubt that a number of infections can be transmitted from patient to dentist during treatment by direct contact, e.g. syphilis, by injection, e.g. hepatitis B, and by aerosols, e.g. the common cold or mumps. Similarly, a dentist may infect a patient during treatment by direct contact, e.g. hepatitis B, and by infective aerosols, e.g. tuberculosis or the common cold. The transmission of hepatitis B from a dentist to his patient probably occurs via infected blood from cuts or abrasions on the dentist's fingers coming into contact with fresh surgical wounds in the patient's mouth. Compared to a dentist, a dental surgery assistant (DSA) is at considerably less risk of either becoming infected or of transmitting infection to a patient during treatment. However, DSAs may become infected by an accidental penetrating wound during handling of contaminated sharp instruments, or by aerosols if working close to the site of operation when high speed instruments are being used. The dental mechanic is at least risk of infection, with the most likely scenario being an accident involving a contaminated small, sharp clasp on a chrome cobalt denture. Finally, it is the responsibility of the dentist to ensure that the transmission of infection from one patient to another does not occur. This is generally achieved by ensuring that all instruments and equipment are rendered safe for use between patients.

The microorganisms which can be transmitted by infected aerosols are shown in Table 17.1. The majority of these are common viral infections. When aerosols are produced by high-speed instruments, two different types of particles are produced. The characteristics of these are shown in Table 17.2 and, while the smaller particles are involved in transmission by inhalation, the larger particles coat surfaces and may enter the tissues via cuts. The microorganisms involved in transmission by direct contact and by injection are hepatitis B virus, human immunodeficiency virus, herpes simplex virus and *Treponema pallidum*. Details about the microorganisms involved in cross-infection in dentistry are given in Chapters 8, 10, 15 and 16.

Table 17.1 Microorganisms transmitted by infected aerosols

Viruses	*Bacteria*
Influenza	*B. pertussis*
Rhinovirus	*M. tuberculosis*
Adenovirus	*Strep. pyogenes*
Mumps virus	
Measles	
Rubella	
Varicella zoster	
Epstein–Barr	

Table 17.2 Characteristics of aerosols produced by high-speed instruments

	Particles	*Droplet nuclei*
Diameter	<100 μm	>100 μm
Time spent airborne	Minutes	Hours
Penetration into respiratory tract	Unlikely	Possible
Likely mode of transmission	Direct contact or in dust	Inhalation

Prevention of cross-infection

Methods of preventing cross-infection can be conveniently considered under two headings: (1) prevention of infection among staff; and (2) prevention of infection from patient to patient. It is also important to obtain a thorough medical history from patients at their initial visit, and, at subsequent visits, to ask specific questions regarding recent medications, current or recurrent illnesses, lymphadenopathy, unintentional weight loss, oral soft-tissue lesions and a past history of jaundice.

Prevention of infection for staff

Infection control for staff can be achieved by using a combination of immunization procedures, protective clothing and, where possible, reducing the level of aerosol production.

Immunization
A list of the vaccines available for health care staff are shown in Table 17.3. Staff who have no natural or acquired immunity to these infections should be immunized by their general medical practitioner.

Table 17.3 Immunization procedures available to health care workers

Disease	Vaccine
Tuberculosis	BCG
Tetanus	Toxoid and 'booster'
Poliomyelitis	Live vaccine
Pertussis	Toxoid
Rubella	Live vaccine
Mumps, measles	Live vaccine
Hepatitis B	Antigen

Protective clothing
The main items of protective clothing required to reduce the risk of cross-infection to dentists and their staff are as follows: (1) a long-sleeved, high-necked clinical coat; (2) rubber gloves which cover cuts and abrasions and may retain a proportion of infective material in an accident with sharp instruments; (3) a face mask to protect the respiratory tract from aerosols generated during the use of ultrasonic scalers or high speed handpieces; and (4) eyeshields to protect the eyes from damage caused by the impact of small fragments of hard tissues on the retina, as well as to prevent possible infection via any small breaks in the conjunctiva. Protection against aerosols can also be achieved by removing the infective particles from the atmosphere at source by employing high-vaccum suction close to the

site where high-speed instruments are being used and by using rubber dams wherever possible. A simple method of reducing the numbers of potentially infectious particles in any aerosol generated during dental treatment is to ask the patient to rinse his mouth with chlorhexidine for three minutes before the use of high speed instruments. Further information concerning the indications for using these protective items is given in Appendix 1. In certain surgical or high-risk situations it may be necessary to wear a surgical gown, hat and overshoes but these are not required for general dental treatment.

Prevention of infection for patients

The main procedures used to ensure that patients do not become infected during dental treatment are as follows: (1) sterilization of instruments and, where necessary, the use of disposable items; (2) a high standard of environmental hygiene; (3) disposal of waste; and (4) where necesssary, the use of protective clothing by dental staff.

Sterilization of instruments
Dentists must know how to render instruments and equipment safe for use in the dental clinic. This does not always require sterility, which means the complete destruction of micro-organisms including bacterial spores. The degree of killing required in any given situation depends on knowledge about (1) the general health of the patient, (2) the nature of treatment required, and (3) the number and types of pathogenic agents likely to be present. Clearly for open heart surgery all instruments and equipment must be sterile. In most dental procedures, the sterility of some instruments and equipment cannot be readily achieved. However as long as approved cleansing and disinfection procedures are performed, this failure to achieve sterility is acceptable.

General principles of sterilization

Generally, moist heat is more efficient in destroying micro-organisms than dry heat, and therefore lower temperatures and times of exposure can be used. It is important that articles should be cleaned before they are sterilized, because the presence of organic matter such as blood or saliva protects bacteria and spores, thus interfering with the sterilization process. Usually a safety margin is built into sterilization procedures, to take account of possible minor inadequacies in the sterilization process.

Autoclaves
A small bench autoclave is the most effective and convenient method of sterilizing instruments in dental practice, the total cycle lasting about 8–10 minutes. The general principle on which autoclaving is based is the fact that when water is heated within a closed container the temperature at which it boils rises above 100 °C. The temperature of the steam which thus forms rises as the atmospheric pressure increases, e.g. at 15 lbs/in^2 the temperature of steam is 121 °C, while at 30 lbs/in^2 a temperature of 134 °C, is achieved. The latter conditions are those incorporated into small bench autoclaves. The higher the temperature reached, the shorter the time required for sterilization to occur; thus at 134 °C only 3–4 minutes is necessary. Steam is more efficient than hot air as a sterilizing agent, partly because it liberates latent heat when it condenses to form water and this heat participates in microbial killing. The simplest type of autoclave is a domestic pressure cooker and in an emergency situation this can be used to sterilize instruments in a dental practice. The small bench autoclaves used in practice do not commonly possess a drying cycle and, therefore, when sterilization is complete the instruments are wet. However, failure to achieve dry instruments is only a drawback when it is intended to sterilize and store instruments in paper-covered packs, which may not be necessary in a busy practice. Recently, bench autoclaves with a drying cycle have become available. The small bench autoclaves are distinct from the high-vacuum machines used in theatre sterile supply units. It is important to test the efficacy of the autoclave at regular intervals, using one of a number of temperature testing kits which are commercially available, e.g. Brownes tubes, or time, steam, temperature (TST) control strips.

Hot air oven
The recommended sterilization cycle for a hot air oven is either 160 °C for 1 hour or 180 °C for 30 minutes. Due to these lengthy processing times, the use of hot air ovens for the routine sterilization of dental instruments is unattractive. It should be remembered that if the shorter of the two cycles is used, the high temperature achieved may damage delicate instruments over a period of time.

260   Cross-infection and sterilization

Hot water boilers
Until relatively recently, the small, electrically-heated boiler was a familiar piece of equipment in many dentists' surgeries. The maximum temperature is 100 °C and will therefore not kill all spores. However, if it is used carefully and correctly, i.e. if the water is changed regularly, the instruments from one patient are not mixed with those from another and a full 15 to 20 minutes exposure is achieved, then all vegetative bacteria and viruses should be destroyed, as long as the instruments were cleaned before being placed in the boiler. However, these and other factors are often difficult to control and overall a hot water boiler is not recommended for rendering instruments safe for clinical use.

Cold 'sterilization'
Disinfectants are chemicals which inactivate vegetative bacteria, viruses and fungi but are barely capable of killing bacterial spores. Many microbiologists consider them to be unreliable, since efficacy usually depends on exposure for lengthy periods of time and their activity can be substantially reduced by the presence of organic matter. However, in dental practice there are some items of equipment which are unlikely to be involved in transmitting infection and cannot easily be sterilized by heat. In this situation disinfection is a satisfactory method of achieving an instrument which is safe to use on another patient. Buffered glutaraldehyde is the only satisfactory disinfectant for use in cold 'sterilization'. Glutaraldehyde kills vegetative microorganisms quickly and bacterial spores very slowly. Although it is not inactivated by organic matter, it does penetrate such material slowly. Direct skin or mucosal contact (e.g. by inhalation of fumes) with the disinfectant should be avoided as it can cause sensitization. Due to such hazards, glutaraldehyde should be used in firmly closed containers and rubber gloves worn when adding and removing instruments. After clinical use, instruments are washed in running water and placed in 2% glutaraldehyde for 10 to 20 minutes. While this time should be sufficient to inactivate all the microorganisms likely to be a problem in cross-infection within the dental surgery, it is not recommended that cold 'sterilization' should be used in place of autoclaving for routine sterilization of dental instruments.

Different methods of satisfactorily rendering dental instruments and equipment safe for use in clinical situations are shown in Table 17.4 and also discussed briefly in Appendix 1.

Table 17.4 Methods of sterilizing dental instruments and equipment

	Autoclave (134 °C 4 min)	Hot air oven (180 °C 30 min)	Cold sterilization (2% Cidex 20 min)	Disposable
Hand instruments	+	+	−	(+)
Syringes	+	+	(+)	−
Needles	(+)	(+)	−	+
Scalpel blades, sutures and needles	(+)	−	−	+
Endodontic instruments	+	(+)	−	−
Orthodontic pliers	+	−	(+)	−
Orthodontic kits for fixed appliances	+	−	(+)	−
Matrix bands	+	+	−	+
Saliva ejectors	+	+	−	+
Cavitron heads*	+	(+)	−	−
Handpieces*	+	(+)	−	−
Steel burs	+	−	(+)	+
Tungsten carbide and diamond burs	+	(+)	(+)	−
3-in-1 syringe tips*	+	−	(+)	+
Tumblers	+	−	−	+

* Essential to follow manufacturers instructions.
+, Recommended method; (+) suitable method in some cases but not recommended routinely; −, unsuitable and not recommended.

Disinfection of dental unit

Between patients, and at the end of clinical sessions, it is important that the surfaces of dental units which have been in direct contact with patients' skin or contaminated by infected waste or instruments, should be cleaned. Generally, dental units have not been designed to enable rapid and effective cleaning to be carried out. The majority of units possess numerous crevices and corners which become contaminated with blood and can be cleaned only with great difficulty. Some recent models are much improved and designed to make cross-infection control as easy as possible. If good clinical operating practice is used by operators, the need for unit disinfection should be mimimal. An operator must remember that once his hands become contaminated with blood or saliva, he must not touch light handles, unit handles, chair controls, radiographs, illuminated viewers, patient notes, pens, X-ray machine cones, etc. With planning and thought preoperatively, these errors in technique can be avoided. If possible, non-operating assistants should perform the above duties. However, if assistants are not available the operator should place a disposable bag over his hand before touching these items. The bag should be subsequently discarded into the infected waste container. Between patients, head rests, bracket tables, dental light handles and chair controls should be disinfected with isopropyl alcohol wipes. For further details of environmental hygiene techniques for use between patients see Appendix 1.

Disposal of waste

It is good clinical practice to separate waste heavily contaminated with blood or saliva from other non-infected waste by placing it directly into a special container, e.g. a plastic bag attached to the unit or a pedal bin with a plastic liner. All disposable sharps, i.e. syringe needles and cartridges, scalpel blades and burs, etc., must be discarded into a rigid, purpose-built container (Figure 17.1). Special care must be taken when disposing of syringe needles to ensure that a needle-stick injury does not occur. A number of special devices to protect staff

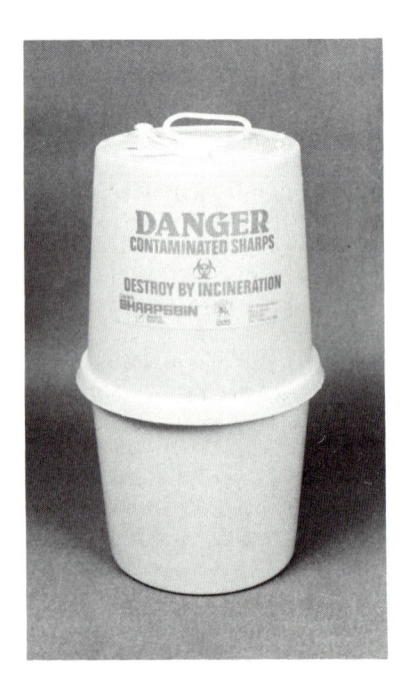

Figure 17.1 A rigid plastic container for the collection of used 'sharps'

against such injuries are now available (Figure 17.2) and one of these should be used. Ideally, infective waste and sharps should be incinerated and individual dentists should arrange for this to be carried out. Infective and domestic waste should not be mixed, nor must the mixture be discarded into domestic dustbins.

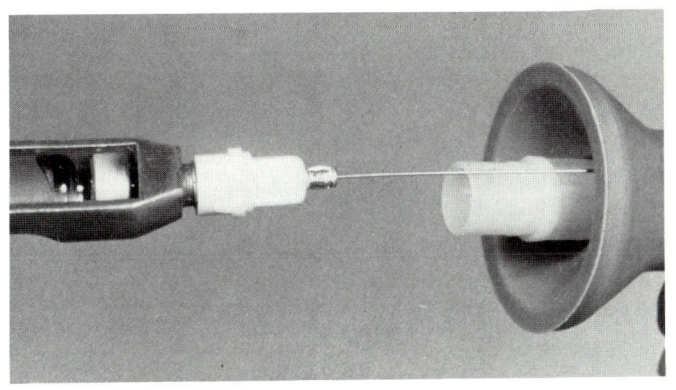

Figure 17.2 The wide rubber rim of the 'aim-safe' device protects against an accidental needle-stick injury which may occur when resheathing a dental syringe needle

Dental treatment of high-risk patients

When a known high-risk patient requires dental treatment, dentists tend to panic and some even refuse to offer treatment. This behaviour is both illogical and reprehensible in a member of a health care profession. It is almost certain that the dentist already treats a number of high-risk patients unwittingly every year, while using the preventive measures which he employs every day on all his patients, without suffering harm. The fact that he knows that a patient belongs to a high-risk group should encourage the dentist to be more careful during treatment but, logically, no extra precautions are required. Additional preventive measures (Table 17.5 and see Appendix 2) may be used during certain forms of treatment, e.g. periodontal surgery. It is important that the dentist does not treat the high-risk patient as a leper, donning the full range of protective clothing and draping the surgery, simply to talk to the patient and carry out a preliminary dental examination. There is little doubt that a combination of knowledge and common sense is essential for a dentist to carry out treatment on risk patients. It should also be remembered that, since aerosols are not important in the transmission of either HBV or HIV, the only chance of infection when treating patients infected with these viruses is by a needle-stick type of injury, where rubber gloves will give little protection. Finally, the ethical and legal implications of refusing to treat a risk patient should be carefully considered by all dentists in the current litigation-conscious society.

Table 17.5 Additional procedures available when treating known high-risk infective patients

Arrange appointment at end of day
Replace protective gown after treatment
Use rubber dam where possible
Use a 3-in-1 syringe with sterilizable tip
Use autoclavable handpiece
Avoid aerosol production as much as possible
Use disposable items where convenient
Use rubber base for impressions and disinfect denture work and impressions prior to handling by dental technician
Protect X-ray film in a sealed plastic bag when in patient's mouth

Further reading

American Dental Association (1988) Infection control recommendations for the dental office and the dental laboratory. *Journal of the American Dental Association*, **116**, 241–248

British Dental Association (1987) *Guide to Blood-Borne Viruses and the Control of Cross-Infection in Dentistry*, 40 pp., BDA, London

Centre for Disease Control (1986) Recommended infection control practices for dentistry. *Morbidity and Mortality Weekly Report*, **35**, 237–242

Samaranayake, L. P. (1987) AIDS, hepatitis B and the general dental practitioner: Part 2 Infection control measures. In *General Dental Treatment* (ed. W. M. Tay), Kluwer, Middlesex, pp. 4.7.2–01–07

Scottish Home and Health Department (1986) *Guidance for Surgeons, Anaesthetists, Dentists and Their Teams in Patients Infected with HTLV III*, Scottish Home and Health Department, Edinburgh

Wood, P. R. (1987) *The Dentist's Guide to Cross-infection Control*, Update–Siebert, Surrey

Appendices

Appendix 1

Code of practice for the use of protective clothing and other procedures associated with the prevention of cross-infection during dental treatment

Protective clothing

Rubber gloves

Rubber gloves of the appropriate type (i.e. sterile gloves for theatre work and non-sterile examination gloves for other procedures) **must always** be worn by dentists and ancillary workers in the following circumstances, where the risks of cross-infection are highest:

1. All oral surgical procedures.
2. All forms of periodontal surgery and during deep and superficial scaling procedures.
3. While participating in the treatment of patients with known infectious diseases, e.g. primary herpetic stomatitis, or recurrent skin lesions, e.g. cold sores, etc.
4. Any other surgical procedures carried out by the dentist where obvious bleeding occurs.
5. When examining all mucosal lesions.
6. Whenever the above-named personnel have obvious cuts, abrasions or dermatological lesions on their hands.

In addition, rubber gloves **should** be worn routinely by dentists and ancillary workers for all other forms of dental examination and treatment, although individuals **should** exercise personal preference in situations which they consider low risk.

Disposable rubber gloves must be worn by dental surgery assistants (DSAs) when assisting in the treatment of patients or handling infected instruments, etc. However, DSAs should wear re-usable, domestic grade, rubber gloves when cleaning instruments prior to sterilization, or dental units.

Ideally, rubber gloves should not be worn when performing procedures which require the use of a naked flame, e.g. a Bunsen, since they may ignite. If rubber gloves require to be worn in these circumstances then extreme care must be taken and cold water used to smother flames if an accident occurs. A newly-designed, electrically-operated Bunsen burner may be useful in resolving this problem.

Protective gloves should be worn for as short a time as possible and removed as soon as patient contact is over. This allows the superficial layers of the skin to dry. After removal of intact gloves, hands should be rinsed thoroughly in water and a handcream applied. This is important to prevent excessive evaporation from the surface of the skin, which leads to chapping. Before leaving the clinic or before replacing damaged gloves, the hands should be washed thoroughly with a disinfectant soap.

There are two views concerning the wearing of rubber gloves during dental treatment: (1) use an inexpensive pair of examination gloves on one patient only and then discard; and (2) use expensive surgical-type gloves, and re-use on 10–12 patients, washing the gloved hands between patients with disinfectant soap. Neither of these methods is ideal. Significant numbers of dentists have difficulty in obtaining examination gloves which fit comfortably, with the result that they may have increased risk of needle-stick-type accidents. In theory, the re-use of gloves after a disinfectant wash should be satisfactory, but much depends on the degree of contamination and the care taken during disinfection. Ideally, a cheap, disposable glove made in a range of sizes would be preferable.

Surgical masks and eye shields

A disposable mask and protective eyeshield must be worn by dentists and ancillary workers when using handpieces and cavitrons. In addition, eyeshields should be worn by the patient during scaling and polishing and four-handed dentistry. Care should be taken to minimize the formation of potentially infectious droplets and aerosols by using rubber dam, where appropriate, and high-volume aspiration.

Rubber gloves, surgical gowns, masks, eyeshields and plastic aprons must not be worn outside clinical areas.

Instruments

Handpieces

A sterile handpiece must be available for the treatment of every patient; this can be achieved if two autoclavable handpieces per dentist are available in each dental practice. If sterilizable handpieces are not in use, this defect should be rectified as soon as possible.

Cavitron heads

A sterile cavitron head must be used for each patient, and the barrel thoroughly wiped with an isopropyl alcohol.

Burs and other attachments to handpieces

Ideally steel burs should be used on only one patient and then discarded into a recommended sharps container. Tungsten carbide and diamond burs, finishing burs, stones and rubber prophylactic cups can be re-used but debris must be removed thoroughly before sterilization by autoclaving, or by cold 'sterilization' if components are heat sensitive. The ideal method of débridement is the use of a powerful small ultrasonic bath with a recommended cleaning solution.

Matrix bands

Matrix bands should be used only once and then discarded into a sharps container or, after thorough cleaning, sterilized with its holder by autoclaving before re-use.

Disposal of sharps

All disposable sharps must be discarded into a recommended sharps container.

Syringes, needles and anaesthetic cartridges

A disposable needle must be used on only one patient before being discarded into a sharps container. Similarly, an anaesthetic cartridge must not be re-used on a second patient. The cartridge must be discarded into a sharps container. A number of devices are available to reduce the potential risk of a needle-stick injury during recapping of a dental syringe needle before discarding into a sharps container. One of these devices should be used (see Figure 17.2).

When conventional disposable plastic syringes are used for venepuncture or injection, no attempt should be made to recap the needle; the syringe with the attached needle should be discarded directly into a sharps container for incineration.

3-in-1 syringes

The barrel of the 3-in-1 syringe must be thoroughly cleaned with isopropyl alcohol between patients, and the tip removed for sterilization by autoclave. If a sterilizable tip is not available, then it must be thoroughly cleaned with alcohol as described above.

Cavitron barrels

Cavitron barrels must be thoroughly cleaned with isopropyl alcohol and the tips autoclaved between patients.

Saliva ejectors

Saliva ejectors must be used either as plastic disposable items or autoclaved between patients.

Cups

Disposable cups should be used and discarded at the end of treatment. Alternatively metal beakers can be sterilized between patients.

Water lines

Each morning, water lines to 3-in-1 syringes and handpieces, etc. must be cleared by running for at least 1 minute before use. If a unit has been out of operation for more than a few days, the time used should be increased to at least 5 minutes.

X-ray equipment and films

Disinfectants should be used at least once a day to control microbial contamination of collimating tubes. Contaminated intraoral X-ray films should be opened using disposable gloves. The film should be dropped out of the packet without touching the film, which can then be processed without contaminating the dark room equipment or solutions.

Disinfection of unit surfaces between patients

If good clinical operating practice is used by operators, the need for unit disinfection should be minimal. Operators must remember that once their hands become contaminated with blood or saliva, they must not touch light handles, unit handles, chair controls, radiographs, illuminated viewers, patient notes, pens, X-ray machine cones, etc. With planning and thought preoperatively, these errors in technique can be avoided. If available, non-operating assistants should perform the above duties. If assistants are not available, the operator should place a disposable plastic bag over his hand before touching the above-named items. The bag should be subsequently discarded into the infected waste container.

Headrests

Must be wiped between patients with isopropyl alcohol. Each day all headrests in use should be cleaned once (or more if necessary) with a general purpose cleaner. Alternatively a disposable plastic cover may be used.

Bracket tables

Should be covered with a disposable paper drape and wiped with isopropyl alcohol between patients. Each day, all bracket tables in use should be cleaned once (or more if necessary) with a general purpose cleaner.

Dental light handles, chair controls, etc.

If touched by the operators, contaminated hands must be wiped with isopropyl alcohol between patients. Each day, all light handles, etc. in use should be cleaned once (or more if necessary) with a general purpose cleaner.

Unit tops

Should be wiped with a general purpose cleaner between patients. For difficult stains, or those contaminated with blood, a hypochlorite cleaner should be used.

Cuspirators

Do not normally require to be disinfected between patients. However at the end of the day a hypochlorite cleaner, household bleach diluted 1:10 to 1:100 (giving 500 to 5000 p.p.m. of chlorine) should be used to clean the surface of cuspirators. Also, if blood or heavy salivary contamination of the unit occurs during the day and cannot be removed by the flushing action of the cuspirator, the same measure should be used.

Sinks in clinical areas

Should be cleaned with a general purpose cleaner at least once a day.

Drainage systems within units

Use cleaners recommended by the manufacturers.

Vomit, etc.

To remove such noxious material from the floor and around the cuspirator, use sodium hypochlorite. In addition, an air freshener should be used in a clinical area.

Cold 'sterilization' of instruments, used in non-invasive techniques, which are sensitive to heat

The only method recommended is 2% glutaraldehyde (Cidex). Wearing domestic-type rubber gloves, wash instruments thoroughly with detergent and hot water; place instruments in 2% glutaraldehyde in a container with a close fitting lid for 30 minutes; rinse items thoroughly with water, and dry before use. The small volumes of in-use glutaraldehyde must be discarded regularly and replaced with fresh disinfectant.

Waste disposal

All waste infected with visible blood or heavily soiled with saliva must be separated in the clinic. It follows that all infected waste must be placed carefully and directly into a container reserved only for infected waste, e.g. plastic bag attached to a unit or a pedal bin with a plastic liner. Non-infected waste must be placed in a separate container or bin.

Handling of impressions and denture work

Impressions and denture work, etc. must be washed thoroughly under running water to remove blood, saliva, etc., and, where applicable, placed in plastic bags before transport to the technology laboratory. Dental technicians should wear disposable rubber gloves when casting impressions but gloves are not necessary after the models have been cast. Technicians must wear disposable rubber gloves when handling or working on denture work which has recently been in contact with blood or saliva. An alginate impression material which has disinfectant properties has recently been produced; this may be useful in cross-infection control.

Appendix 2

Additional procedures available when treating known high-risk infective patients

Infective high-risk patients

1. Patients who have serum hepatitis.
2. Patients who are both HBsAg and HBeAg positive.
3. Patients who are HBsAg positive but have unknown 'e' antigen status.
4. Patients with confirmed AIDS.
5. Patients with persistent generalized lymphadenopathy (PGL).
6. Patients with AIDS-related complex (ARC).
7. Patients who have antibodies to HIV.
8. Patients with open tuberculosis.
9. Patients requiring emergency dental treatment and whose history suggests infective risk status (serological tests must always be carried out on these patients to confirm or deny their suspected infective status).

Instruments, materials and disinfection

1. High-speed handpieces and cavitrons may be used if high volume suction is used during the operation.
2. A 3-in-1 syringe may be used if the working tip can be sterilized after use. If this is not possible a disposable syringe and water should be used for washing debris from cavities and cotton wool pledgets for subsequent drying of cavities.
3. If an experienced operator is able to apply well-adapted rubber dam to the operating site, high-speed handpieces and a non-sterilizable 3-in-1 syringe may be used. The intraoral operating area should be cleaned with dilute sodium hypochlorite (Milton) before starting.
4. For impressions, disposable impression trays and rubber-based, or another impression material which remains stable when treated with glutaraldehyde, should be used.
5. Disposable paper drapes should be used to cover bracket

tables and side benches. If heavy contamination with blood is likely, a water-impermeable covering should be placed beneath the paper drape.
6. All working surfaces, including the headrest, cuspirator bowl and unit light, should be disinfected with hypochlorite if no metal surfaces are present, otherwise 2% glutaraldehyde should be used.
7. Thoroughly wash the spittoon and surrounding area with 2% glutaraldehyde. Flush 2% glutaraldehyde through the spittoon and saliva ejector systems for 2 minutes. After 3 hours treat the spittoon system with the cleaner recommended by the manufacturer.

Radiography

1. The radiographer must wear rubber gloves.
2. Intraoral films should be placed in a sealed plastic bag with the patient holding the film *in situ* while radiographs are being exposed. The cone should be covered with a disposable plastic bag.
3. Once the film has been exposed the gloved radiographer must be responsible for opening the sealed plastic bag and allowing the film to be deposited in a clean container before being processed.

Dental technicians

1. Clinicians must give technicians advance warning of impressions, denture work, etc. from high-risk patients.
2. Impressions or denture work should be placed in a plastic bag containing 2% glutaraldehyde and decontaminated for one hour. After decontamination the impressions or denture work are removed by the DSA, rinsed and transferred to a fresh solution of glutaraldehyde; they are then sent to the dental technology laboratory.
3. The dental technician will ensure that the impression or denture work suspended in glutaraldehyde is stored at room temperature for 3 hours or overnight (whichever is more convenient), before work is started.
4. The next morning the technician, wearing disposable rubber gloves, removes the impression from the plastic bag and thoroughly rinses it in water. The gloves, which are a

protection from glutaraldehyde (which may irritate the skin), are discarded and the models cast as normal. No further precautions to prevent cross-infection are necessary during subsequent technical work.

Index

Numbers in italics indicate an illustration or table.

Abscess
  dentoalveolar, 75–82
  periodontal, 83–85
  sample collection, 191–192
Acetate production, 43
Actinobacillus spp
  actinomycetemcomitans
    in localized juvenile periodontitis, 68
    pathogenic factors produced by, 66
    in refractory chronic periodontitis, 66–67
  characteristics, *13*
Actinomyces spp
  in actinomycosis, 90–91
  characteristics, *10*
  viscosus role in caries, 41–42
Actinomycosis
  cervicofacial, 89–92
  salivary gland, 165–166
Acyclovir, 219–220
  for chickenpox, 149
  for herpes virus, 147
  for shingles, 151
Adenovirus structure, *140*
Aerosols
  microorganism transmission by, 256
  protection against, 257–258
AIDS
  candidosis and, 124, 134, *136*
  definition, 239
  diagnosis, 252
  epidemiology, 240
  management, 252
  natural history and clinical features, 243–245
  oral manifestations *179*, 245–250

AIDS (*cont.*)
  staff with, 252–253
  transmission, 250–251
Amoxycillin, 214
  for dentoalveolar abscess, 81
  prophylaxis, 174–176
Amphotericin B, 219
Ampicillin, 214
  activity spectrum, *207*
Angina, Ludwig's, 82–83
Angular cheilitis, 131–134
Antibiotics
  for craniofacial actinomycosis, 92
  for dentoalveolar abscess, 81
  for gonorrhoea, 107
  for infective endocarditis, 172–174
  for parotitis, 163–164
  prophylaxis, 174–177
  sensitivity testing, 201–203
  sore tongue from, 126–127
  for staphylococcal submandibular lymphadenitis, 89
  for suppurative osteomyelitis of jaw, 88
  susceptibility of organism to, 201
  for syphilis, 112
  for xerostomia, 183
Antibodies to periodontal microorganisms, 55–56
Antifungal agents, 218–219
  for acute atrophic candidosis, 127
  for angular cheilitis, 133
  for chronic candidosis, 137
    atrophic, 131
    hyperplastic, 129
  prophylaxis, 180
  for thrush, 126
Antimicrobial agents, 47–48, 204–222
  activity spectrum, 206–207
  bacteriostatic and bactericidal, 204–205

277

278 Index

Antimicrobial agents (*cont.*)
  choice, 206
  combination, 207
  in endodontics, 98–100
  failure, 211
  laboratory investigations for, 200–203
  mode of action, 205
  for periodontitis, 70
  pharmacodynamics, 209–211
  prescribing, 208–209
  prophylaxis, 207–208
  use in dentistry, 211–212
  see also specific types and agents
Antiseptic prophylaxis, 174
Antituberculous therapy, 115
Antiviral agents, 219–220
Arachnia characteristics, *11*
Aspergillosis, *138*
Autoclaves, 225

Bacteraemia
  causes, 171–172, 173
  prophylaxis, 174
Bacterial infection
  AIDS and, 248
  endocarditis, 167–174
  oral tissues, 103–121
  salivary gland, 161–164
  specimens, 191–194
Bacterionema spp characteristics, *11*
Bacteroides spp
  characteristics, *15*
  in chronic periodontitis, 65, *66*
  in dentoalveolar abscess, 79, *80*
  gingivalis in chronic periodontitis, 65, *66*
Bifidobacterium spp characteristics, *11*
Blastomycosis, South American, 137–138
Bone infection, tuberculous, 114
Boilers, hot water, 260
Borrelia vincenti causing tonsillitis/pharyngitis, *104*
Branhamella spp characteristics, *12*
Bur sterility, 269

Calculus role in periodontal disease, 59–60
Cancrum oris, 73
Candida spp, 123–124
  albicans, 123
  characteristics, *16*
  endocrinopathy syndrome, 135, *136*

Candidosis, oral, 123–124
  acute atrophic, 126–127
  pseudomembranous, 124–126
  AIDS and, 246
  chronic atrophic, 129–131
  hyperplastic, 127–129
  multifocal, 134
  factors predisposing to, *122*
  following xerostomia, 183
  in immunocompromised patients, *179*
  management, *181*
  mucocutaneous, 134–137
  diffuse chronic, 135, *136*
  familial chronic, 135, *136*
  sampling and examination, 198–200
Capnocytophaga spp
  characteristics, *13*
  in chronic periodontitis, 66
  in localized juvenile periodontitis, 67
Carbohydrates
  caries and, 38
  salivary levels, 42–43
Carcinoma
  AIDS and, 249
  viruses and, 156
Caries
  clinical presentation and diagnosis, 35–36
  in dentine causing pulp infection, 95
  following xerostomia, 182–183
  in immunocompromised patients, *179, 181*
  management, 42–46
  microbiological tests in, 44–45
  microbiology of, 35–40
  prevention, 46–50
Cartridge disposal, 269
Cavitron barrel sterility, 270
Cephalosporins, 215
  activity spectrum, *207*
  interaction, *211*
Cephradine, 215
Cervicofacial actinomycosis
  aetiology, 89–90
  clinical presentation, 90
  microorganisms, 90–91
  pus sampling, 91–92
  treatment, 92
Chancre, 108
Cheilitis, 131–134
Cheilocandidosis, 134
Chickenpox, *141*, 148–149

Chlorhexidine mouth wash, 47–48
Clavulanic acid, 214
Clindamycin, 216
  for dentoalveolar abscess, 81
Clostridium tetani examination, 119
Clothing, protective, 257–258, 267–268
Cloxacillin, 215
Coccidioidomycosis, *138*
Corynebacterium diphtheriae causing tonsillitis/pharyngitis, *104*
Cotrimoxazole, 217–218
Coxsackie virus, *141*, 152–154
  culture, *196*
Crevicular fluid
  effects on microbial ecology, 22–24
  periodontal disease and, 54
Cross-infection, 255–264
  dental technician, 275–276
  prevention, 256–264, 267–276
  route of, 255
Cryptococcosis, *138*
Culture
  Mycobacterium spp, 114–115, 118
  Neisseria spp, 106–107
  specimen, 18, 189–191
Cups, disposal of, 270
Cuspirator cleaning, 271–272
Cysts, periapical, 95
Cytomegalovirus, 141, 152
  culture, *196*
  salivary gland, 160–161
Cytotoxic therapy, immunodeficiency and, 178, *179*

Dapsone therapy, 118
Dental unit cleaning, 261–262, 271
Dentoalveolar abscess
  aetiology, 75–77
  clinical presentation, 78
  infection spread and sequelae, 77–78, *82*
  microorganisms involved, 78–79, *80*
  sample collection, 81, 191–192
  treatment, 81
Dentures, inducing stomatitis, 129–131
Dermatitis, herpetic, 146
Di George's syndrome, *136*
Dietary factors in dental caries, 38
Dip-slide test, 45
Disc diffusion tests, 201–202
Disinfection, 260, 270–272, 274–275
Dry mouth, *see* Xerostomia

Eczema herpeticum, *141*, 147
Eikenella spp
  characteristics, *13*
  corrodens in periodontitis, 66
Enamel structure, plaque formation and, 37
Endocarditis, infective, 167–174
Endocrinopathy, candidal, 135, *136*
Endodontic therapy, microbiology in, 92–100
Entamoeba spp, *16*
Epstein-Barr virus, 151
  AIDS and, 248
Erythromycin, 215–216
  activity spectrum, *207*
  for dentoalveolar abscess, 81
  interaction, *211*
  prophylaxis, 174
Eubacterium spp characteristics, *11*
Eye shields, 268

Fissure sealants for caries prevention, 49–50
Flora, *see* Microorganisms
Flucloxacillin, 215
Fluoride for caries prevention, 47, 48
Fungal infections, 122–139
  sampling and examination, 198–200
Fusidic acid, 218
Fusobacterium spp
  characteristics, *14*
  in acute ulcerative gingivitis, 71–72
  in dentoalveolar abscess, 79, *80*
  nucleatum causing tonsillitis/pharyngitis, *104*

Geotrichosis, *138*
Gingivitis
  acute ulcerative, 70–73
    AIDS and, 248
    sampling, 193–194
  chronic, 60–62
    clinical presentation, 60–62
    microorganisms in, *58*, 62
    transition to periodontitis, 62–63
    treatment, 62
Gingivostomatitis, *141*
Glossitis, atrophic, 110
Glossodynia, 127
Gloves, rubber, 267–268
Glucose metabolism, 43
Glucosyltransferase vaccines, 48–49
Glutaraldehyde, 260, 272

Gonorrhoea, 103–107
  salivary gland, 165
Granuloma, periapical, 94–95
  tuberculous, 114
Granulomatous disease, chronic, 136
Gumma, 109–110

Haemophilus spp characteristics 13
Hairy leukoplakia, AIDS and, 247–248
Hand foot and mouth disease, 141, 153
Handpiece sterility, 268–269
Head rest cleaning, 271
Heart
  disease leading to infective endocarditis, 168
  valve prosthesis patients, 177
Hepatitis, 223–238
  A, 225–226
  B in antigen positive staff, 233
    carrier state and identification, 228–229
    dentists and, 231
    identification and treatment of patients with, 232–233
    immunization, 234–235
    inoculation injury, 235–236
    prophylaxis, 233–235
    risk of transmission, 251
    serological markers and microbiological diagnosis, 229–230
    terminology, 227
  delta, 236–237
  non-A non-B, 236
Herpangina, 141, 154
Herpes virus, 143–151
  in immunocompromised patients, 179
  labialis, 145
  management, 181
  simplex, 141
    AIDS and, 246
    culture, 196
    primary, 143–144
    secondary, 144–147
  structure, 140
  therapy, 219–220
  zoster, *see* Varicella zoster
Hip joint prosthesis patients, 177
Histoplasmosis, 137–139
HIV, 240–243
  control, 251

HIV (*cont.*)
  diagnosis, 252
  management, 252–253
  in saliva, 249–250
  in staff, 252–253
  structure, 140
  transmission, 250–251
HTLV, 241
Hutchinson's incisors, 111
Hygiene measurements, 261–262, 264, 267–276
  hepatitis risk and, 233, 235–236
Hyperimmuneglobin immunization, 234
Hypersensitivity mechanisms in periodontitis, 55–56, 57

Idoxuridine, 219
Immunization
  for caries prevention, 48–49
  for hepatitis prevention, 234
  of staff, 257
Immunodeficiency, 77–83
  candidosis in, 135–137
  causes, 177–178
  herpes simplex in, 144
  see also AIDS
Immunofluorescence techniques, 197–198
Impression hygiene, 272–273, 275
Imprint sampling, 200
Infection
  dentoalveolar, 75–92
  source in endodontics, 93
  spread from periodontal abscess, 83–84
Infectious mononucleosis, 151–152
Infective endocarditis, 167–174
  pathogenesis, 168–171
  prophylaxis, 172–174, 175
Infective high risk patients, 274
Inoculation injury, 235–236
Instrument
  cross-infection prevention and, 268–270, 274–275
  sterilization, 258–261, 272
Interferon action in viral infection, 142

Jaw, suppurative osteomyelitis of, 85–88

Kaposi's sarcoma in AIDS, 249
  varicellform eruptions, 141, 147

Koplik's spots, 155

Laboratory
 analysis, 189–191
 investigations, 194–203
 reports, 187–191
Lactate production, 43
Lactobacillus spp
 assessment for, 45
 role in caries, 40–41
 sampling, *10*
Leprosy, 115–118
Leptotrichia spp characteristics, *14*
Leukaemia, oral lesions in, *179*
Leukoplakia
 candida, 127–129
 hairy, AIDS and, 247
Lincomycin, 216
Ludwig's angina, 82–83
Lymphadenitis
 submandibular, 88–89
 tuberculous, 113–114
Lymphadenopathy, AIDS and, 248
Lymphocytes, T, 143
 HIV affinity for, 239, 242

Masks, 268
Matrix band sterility, 269
Measles, *141*, 155–156
Methicillin, 215
Metronidazole, 217
 activity spectrum, *207*
 for acute ulcerative gingivitis, 73
 for dentoalveolar abscess, 81
 interactions, *211*
Miconazole, 219
Micrococcus spp characteristics, *12*
Microorganisms
 assessment in caries, 44–46
 in bacterial parotitis, 162–163, 164–165
 in chronic gingivitis, 62
 in chronic periodontitis, 63–67
 in dentoalveolar abscess, 78–79, *80*
 in endodontic infection, 95
 fusidic acid-sensitive, 218
 in infective endocarditis, 169–171
 lincomycin/clindmaycin-sensitive, 216
 in localized juvenile periodontitis, 67–68
 in Ludwig's angina, 82–83
 metronidazole-sensitive, 217

Microorganisms (*cont.*)
 normal oral, 7–20
  acquisition, 26–27
  atmospheric conditions, 18–19
  culture, 18
  dispersion, 17
  ecology, adherence and metabolism, 25–26
   anatomical factors, 22
   crevicular fluid, 22–24
    microbiological factors, 24
    miscellaneous factors, 24–25
  enumerating techniques, 19
  incubation, 18
  interpretation of results, 19
  sample transport, 17
  sites, 21
 penicillin-sensitive, 213, 214
 in periodontal abscess, 84
 phagocytosis, 54–55
 in plaque, 27–31, 38–42
  control of 47–49
  in gingival disease, *58*
  role in periodontal disease, 56–59
 in prepubertal periodontitis, 68–69
 pulp invasion by, 75–77
 in rapidly progressive periodontitis, 69
 role in acute ulcerative gingivitis, 71–72
 in suppurative osteomyelitis of jaw, 87
 tetracycline-sensitive, 218
 transmission by aerosols, 256
 in xerostomia, 182
 see also Caries; Plaque
Minimum bactericidal concentration
 assessment, 202–203
Minimum inhibitory concentration
 assessment, 202–203
Mouthwash prophylaxis, 180
Mucormycosis, *138*
Mucosal lesions
 in candidosis, 124–126
  acute candidosis, 127
  chronic atrophic, 130
  chronic hyperplastic, 127–128
 in chickenpox, 148–149
 in gonococcal infection, 105
 in hand foot and mouth disease, 153
 in herpangina, 154
 in immunocompromised patients, 179–180
 in infectious mononucleosis, 152

Mucosal lesions (*cont.*)
  in measles, 155
  sampling, 8, 193
  in shingles, 150
  in syphilis, 109
  in tuberculosis, 113
Mucositis, *179, 181*
Mulberry molars, 110–111
Mumps virus, 141, 159–160
  culture, *196*
Mycobacterium spp
  leprae, 115–116
    examination, 118
  tuberculosis, 112
    examination, 114–115
Mycoplasma spp characteristics, *16*
Mycoses, systemic, 137–139

Needle disposal, 269
Neisseria spp
  characteristics, *12*
  gonorrhoeae, 105
    examination, 106–107
  in salivary gland, 165
Noma, 73, 156
Nutrient sources for microbial growth, 26
Nystatin, 218–219

Ocular herpes, 147
Oral rinse sampling, 199–200
Oropharyngeal gonorrhoea, 106
Osteomyelitis, *179, 181*
  of jaw, suppurative
    aetiology, 85–86
    clinical presentation, 86
    laboratory investigation, 87
    microorganism source, 87
    treatment, 88
  syphilitic, 110
Ovens, hot air, 259
'Owl-eye' inclusion bodies, 151
Oxidation-reduction level in oral sites, 24–25

Papillary atrophy, 110
Papilloma virus, AIDS and, 247–248
Papovavirus, 140–141
Paracoccidioidomycosis, 137–138
Paramyxovirus, *141*
  structure, *140*, 155–156, 159
Parotid gland infection
  gonococcal, 106
  pus collection, 192–193

Parotitis
  acute suppurative, 161–164
  endemic, 159–160
  recurrent, 164–165
Pellicle formation in plaque, 28
Penicillin, 212–215
  activity spectrum, 206, *207*
  benzyl, 213
  broad-spectrum, 214
  for dentoalveolar abscess, 81
  depot, 213–214
  for gonorrhoea, 107
  interactions, *211*
  isoxazolyl, 215
  penicillinase-resistant, *207*
  phenoxymethyl, 213
  for prepubertal periodontitis, 68
  procaine, 213
  prophylaxis, 174
  types, *212*
Periapical infections pathogenesis, 93–95
Periodontal
  abscess, 83–85
  disease, factors involved in, 52–60
    gingivitis progressing to, 60–63
    in immunocompromised patients, *179, 181*
  sampling, 193–194
  types, 51, *52*
Periodontitis
  chronic, 63–67
    development of, *61*
    microorganisms in, *58*
    refractory, 66–67
  localized juvenile, 67–68
    microorganisms in, *58*
    sampling, 194
  prepubertal, 68
  rapidly progressive, 68–69
    microorganisms in, *58*
Periodontium
  healthy, *53*
  marginal, defence mechanisms in, *54*
Phagocytosis, 54–55
  viral, 142
Pharyngitis, 103, *104*
  gonococcal, 105
Plaque, 27–31
  bacterial deposition, 29
  calcified, *see* Calculus
  formation, 28
  metabolism, 42–44

Plaque (*cont.*)
  microorganisms in, 38–42
  assessment, 45–46
  control, 47–49
  removal aids, caries prevention and, 47
  role in periodontal disease, 56–59
  sampling, 8, 17
  structure, 30–31
  subgingival, causing periodontal disease, 52, 53
Polymorphonuclear leucocytes, bacterial phagocytosis by, 54–55
Poxvirus structure, *140*
Propionibacteria spp characteristics, *11*
Prosthesis patients, 177
Pulp infection
  host defence mechanism, 95
  microbiology, 75–77, 95–96
  pathogenesis, 93–95
Pus
  collection, 81, 84, 87, 91–92, 191–193
    from parotid gland, 162
    granular, 89
    spread, 77–78, 83–84
Pyruvate degradation, 43

Radiography hygiene, 275
Radiotherapy, immunodeficiency and, 178, *179*
Redox potential in oral sites, 25
Retroviruses, 241
Risus sardonicus, 120
RNA viruses, 142, 159, 241–243
Root canal
  microbiological assessment, 97–98
  sampling, 98
  treatment, 98–100
Rothia spp characteristics, *11*

Saliva
  effects on microbial ecology, 22, *23*
  ejector sterility, 270
  flow, plaque formation and, 37
  HIV in, 249–250
  microorganism assessment in, 45
  problems in xerostomia, 182–183
  sampling, 8
Salivary gland infection, 158–166
Selenomonas spp characteristics, *15*
Serodiagnosis, viral infection, 196–198
Sharps disposal, 262–263, 269

Shingles, *141*, 149–150
Sialadenitis, *179*, *181*
  bacterial, 161–164
  submandibular, 165
  viral, 159–161
Simonsiella spp characteristics, *14*
Sjögren's syndrome, 182–183
  caries in, 37–38
Skin lesion sampling, 193
Specimen collection and transport, 187–189
  bacterial, 191–194
  fungal, 198–200
  pus, 81, 84, 87, 91–92
  root canal, 98
  virological, 194–198
Spirochaete spp in chronic periodontitis, 66–67
Sporotrichosis, *138*
Staphylococcus spp
  aureus causing angular cheilitis, 132–133
  in parotitis, 163
  resistance, 215
  causing submandibular lymphadenitis, 88–89
  in suppurative osteomyelitis of jaw, 87
Sterility testing in endodontics, 97
Sterilization, 258–261, 272
Stomatitis
  denture-induced, 129–131
  gangrenous, 156
  gonorrhoeal, 105
  primary herpetic, 143–144
Streptococcus spp
  acquisition, 27
  characteristics, *9–10*
  in infective endocarditis, 171
  mutans, antigens, 48–49
    assessment, 44–45
    role in caries, 39–40
  pyogenes causing tonsillitis/pharyngitis, 103, *104*
Submandibular
  lymphadenitis, 88–89
  sialadenitis, 165
Sucrose, caries and, 38
Sugar substitutes for caries prevention, 46–47
Sulphonamides, 217–218
Sulphur granules, 89, 91–92
Syphilis
  congenital, 110–111

Syphilis (cont.)
  laboratory diagnosis, 111–112
  late and quaternary, 110
  latent and tertiary, 109–110
  primary, 107–108
  secondary acquisition, 108–109
  treatment, 112
Syringe
  disposal, 269
  sterility, 270

Technician hygiene, 275–276
Tetanus, 119–121
Tetracycline, 216–217
  activity spectrum, *207*
  for gonorrhoea, 107
  interaction, *211*
  for periodontitis, 68, 70
Thrombus formation in infective endocarditis, 168–169
Thrush, 124–126
Tongue lesions
  in acute atrophic candidosis, 126–127
  in leprosy, 118
Tonsillitis, 103, *104*
Toothbrushing, caries and, 47
Transport media, 188–189
  viral, 194–195
Treponema
  characteristics, *15*
  pallidum, 107
    examination, 111–112
    salivary gland, 165
Trichomonas spp characteristics, *16*
Triplopen, 214
Tropism, tissue, 140–141

Tuberculoid leprosy, 116
Tuberculosis, 112–115
  salivary gland, 165–166

Ulcer
  snail track, 109
  tuberculous, 113

Varicella zoster virus, *141*, 148–151
  culture, *196*
Veillonella spp
  characteristics, *12*
  role in caries, 42
Viral infections, 140–157
  AIDS and, 246
  hepatitis, 223–238
    aetiology, *223*
    clinical features, *224*
    epidemiology, *224*, 227
  recovery and immunity to, 142–143
  salivary gland, 159–161
  sampling and examination, 194–198
Viruses
  interaction with host cells, 141–143
Vomit, clearing up, 272

Waste disposal, 262–263, 272
Water line disinfection, 270
White spot lesion diagnosis, 35–36
Whitlow, herpetic, 146
Wolinella spp
  characteristics, *15*
  in chronic periodontitis, 66

Xerostomia, *179*, *181*, 182
  sequelae, 166, 182–183
X-ray equipment disinfection, 270, 275